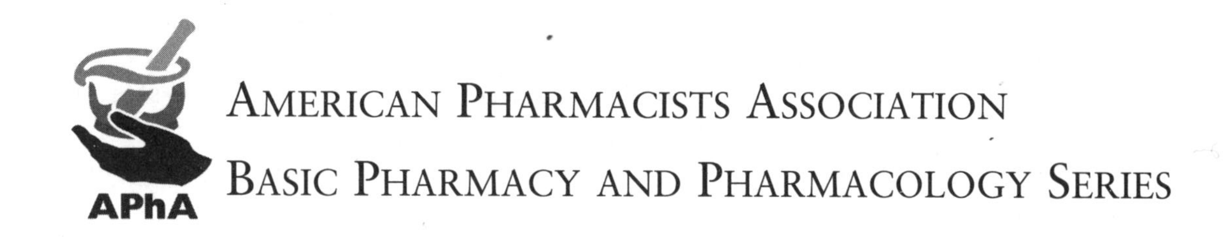

AMERICAN PHARMACISTS ASSOCIATION
BASIC PHARMACY AND PHARMACOLOGY SERIES

The Pharmacy Technician

WORKBOOK AND CERTIFICATION REVIEW

-SIXTH EDITION-

PERSPECTIVE PRESS
MORTON PUBLISHING COMPANY
www.morton-pub.com

Morton Publishing

Printed in the United States of America.

Morton Publishing Company
925 West Kenyon Avenue, Unit 12
Englewood, CO 80110
phone: 1-303-761-4805
fax: 1-303-762-9923
www.morton-pub.com

International Standard Book Number

ISBN 13: 978-1-61731-488-9

10 9 8 7 6 5 4 3

Cover design by Brenda Carmichael; cover photo copyright catalin eremia/Shutterstock.com

For Joanie and Hannah

NOTICE

To the best of the Publisher's knowledge, the information presented in this book follows general practice as well as federal and state regulations and guidelines. However, please note that you are responsible for following your employer's and your state's policies and guidelines.

The job description for pharmacy technicians varies by institution and state. Your employer and state can provide you with the most recent regulations, guidelines, and practices that apply to your work.

The Publisher of this book disclaims any responsibility whatsoever for any injuries, damages, or other conditions that result from your practice of the skills described in this book for any reason whatsoever.

The Pharmacy Technician Workbook and Certification Review

Sixth Edition

Table of Contents

TABLE OF CONTENTS

TABLE OF CONTENTS

Preface

This Workbook

This workbook was developed to correspond with the textbook, ***The Pharmacy Technician***, Sixth Edition, by Perspective Press. For pharmacy technician students, it is a valuable tool for success. It provides a useful format for memorizing important information and checking knowledge. Hundreds of key concepts and terms are carefully explained, and there are over 1,000 exercises and problems for self testing. Answers are included at the back of the book

"In the Workplace" sections provide sample job-related documents and skills checklists. The sixth edition includes additional skills checklists as well as "In the Workplace Activities" that provide practical, active-learning applications of the concepts discussed in *The Pharmacy Technician.* Working out these exercises and activities successfully will help you to succeed in your training.

It is important for you follow through the workbook chapter by chapter and try answering the questions before looking up the answers. Once you have completed a chapter, review it and try to memorize the answers to the questions you missed.

Fifteen labs follow the chapters. The labs are organized by drug classifications. Each lab consists of a drug card activity, a series of web-research activities, and, starting with the fourth lab, prescription label preparation activities. The first three labs also provide foundational activities that support concepts presented in Chapters 1–3 of *The Pharmacy Technician.*

An Exam Review Guide

This workbook can also be used as a review guide in preparing for the national Pharmacy Technician Certification Examination (PTCE) given by the Pharmacy Technician Certification Board (PTCB) or the Exam for the Certification of Pharmacy Technicians (ExCPT) given by the National Healthcareer Association (NHA). All its chapters are important in taking either certification exam. However, special attention should be placed on Chapter 6—Calculations, because the both exams will have calculation-type problem solving that is often challenging for technicians taking either exam. The method used (ratio and proportion) in this section will solve any calculation problem you come across on the certification exams as well as most problems in the pharmacy setting. A careful review of this workbook will prepare you for much of either certification exam. However, some questions on the exams require knowledge gained from practice as a technician. Pharmacy technicians who have work experience in a pharmacy setting will therefore have an advantage in taking a certification exam. As an additional study tool, we have included two practice exams at the back of this book, one in the format of the PTCE and one in the format of ExCPT.

Overview: Pharmacy Technician Certification Exam (PTCE)

The national Pharmacy Technician Certification Examination was established to allow the certification of technicians. The need for highly qualified pharmacy technicians is increasingly important because pharmacists are relinquishing many dispensing duties for more clinical ones, and the technician is playing a greater role.

The PTCE is given by the Pharmacy Technician Certification Board (PTCB) and applications are accepted continuously throughout the year. Applications are submitted online and applicants must take the exam within 90 days of submitting their application. The time frame may be less if a sponsor is paying the exam fee. The cost of taking this examination is, at the time of this, writing $129.

THE PTCE EXAM

The PTCE contains 10 multiple choice pre-test questions and 80 multiple choice exam questions. The multiple choice format has four possible answers with only one answer being the best or most correct.

The time limit for taking this examination is two hours: One hour and 50 minutes are for answering test questions and 10 minutes are for survey questions and a tutorial. Exams are given in a computer-based testing (CBT) format and each candidate is presented with different items.

SCORING OF THE PTCE EXAM

The scoring of the examination is based on the combined average of scores in nine knowledge domains:

1. Pharmacology for Technicians (13.75%)
2. Pharmacy Law and Regulations (12.5%)
3. Sterile and Nonsterile Compounding (8.75%)
4. Medication Safety (12.5%)
5. Pharmacy Quality Assurance (7.5%)
6. Medication Order Entry and Fill Process (17.5%)
7. Pharmacy Inventory Management (8.75%)
8 Pharmacy Billing and Reimbursement (8.75%)
9. Pharmacy Information Systems Usage and Application (10%)

The combined score range of the exam is 300 to 900 points with 650 points required to pass.

WHAT YOU NEED TO KNOW FOR THE PTCE EXAM

Specific information on examination content is provided in the PTCB's *Guidebook to Certification,* which can be downloaded from their website: www.ptcb.org. For additional information, contact the PTCB at:

Pharmacy Technician Certification Board
2215 Constitution Avenue, NW
Washington, DC 20037-2985
800-363-8012 (phone)
202-888-1699 (fax)
e-mail: contact@ptcb.org
www.ptcb.org

OVERVIEW: EXAM FOR THE CERTIFICATION OF PHARMACY TECHNICIANS (ExCPT)

The ExCPT is sponsored by the National Healthcareer Association (NHA). The mission of the NHA is "empowering people to access a better future."

The exam is offered throughout the year at PSI testing sites. Candidates may register online at http://www.nhanow.com. The cost of taking this examination is, at the time of this writing, $105.

PREFACE

THE ExCPT EXAM

The ExCPT exam blueprint specifies that the exam contains 20 multiple choice pre-test questions and 100 scored multiple choice questions. The multiple choice format involves four possible answers with only one answer being the best or most correct. The time limit for taking this examination is two hours and 10 minutes. Exams are given in a CBT format and each candidate is presented with different items.

SCORING OF THE ExCPT EXAM

ExCPT exam questions fall in three main areas:

1. Regulations and Pharmacy Duties (~35% of exam)
2. Drugs and Drug Therapy (~11% of exam)
3. Dispensing Process (~54% of Exam)

Scoring for the ExCPT exam is established by the NHA ExCPT Expert Panel. The passing score is a scaled score of 390 out of 500.

WHAT YOU NEED TO KNOW FOR THE ExCPT EXAM

Specific information on examination content is provided in the ExCPT *Candidate's Guide*, which is available from www.nhanow.com. For additional information, contact the NHA at:

National Healthcareer Association
11161 Overbrook Road
Leawood, KS 66211
800-499-9092 (phone)
913-661-6291 (fax)
e-mail: info@nhanow.com
http://www.nhanow.com

ACKNOWLEDGEMENTS

Mary F. Powers, Ph.D., R.Ph., Professor and Associate Dean, University of Toledo College of Pharmacy & Pharmaceutical Sciences

Mary Powers has been a key contributor on ***The Pharmacy Technician***, on which this workbook is based, since the first edition. She was a key contributor to the first edition of ***The Pharmacy Technician Workbook and Certification Review*** and has been its sole author since the second edition. For the sixth edition of the workbook Mary collaborated with the publisher on a plan for the new edition, updated existing content to reflect changes made in the corresponding text, updated the PTCE and ExCPT Practice Exams, provided the appendix on the Top 200 Most-Prescribed Drugs by Classification, and developed over 70 new In the Workplace Activities, many of which are based on the ASHP Model Curriculum. Mary also tirelessly provided feedback and guidance throughout the development of both texts. Her input has always been enormously helpful, and we cannot thank her enough.

OTHERS WE'D LIKE TO THANK

We would also like to acknowledge Joe Medina, CPhT, B.S. Pharmacy for his contribution to the first edition of the workbook by writing many of the exam review questions and other exercises. Joe has served as the Chairperson/Program Director for the Pharmacy Technician Programs at Front Range Community College and Arapahoe Community College.

In addition, this workbook would not have been possible without the efforts of a large number of people who also worked on the corresponding text, ***The Pharmacy Technician***, Sixth Edition. We would like to thank them again for their contributions: Robert P. Shrewsbury, Ph.D., R.Ph., Associate Professor, Division of Pharmacy Advancement and Clinical Education, Eshelman School of Pharmacy, University of North Carolina-Chapel Hill, and member of the USP Expert Committee on Compounding; Cindy B. Johnson, R.Ph., M.S.W., and Preceptor with the Colorado State Board of Pharmacy; Brenda Vonderau, B.Sc. (Pharm.), Director, Clinical Products, OptumRX; Steve Johnson, B.S., R.Ph.; Ruth Gilman, Pharm.D., Pharmacist, CVS Pharmacy; Pamela Nicoski-Lenaghan, Pharm.D., BCPS, Clinical Pharmacist, Loyola University Medical Center; and Britta Young, Pharm.D., Clinical Pharmacist, Ann & Robert H. Lurie Children's Hospital of Chicago.

We are also grateful to the following reviewers who provided feedback on the fifth edition, which we used to help prepare the sixth edition: Nicole K. Daw, CPhT, RPT, Lincoln Technical Institute, Fern Park, FL.; Roberta M. Ebbers, CPhT, M.A., National American University, Bloomington, MN; Steven L. Feaver, CPhT, PIMA Medical Institute, Las Vegas, NV; Amber Fowler, CPhT, Pima Medical Institute, Mesa, AZ; Valerie Greene, B.S., St. Louis College of Health; Heather Krehl, B.S., PhTR, CPhT, Milan Institute, Amarillo, TX; Jaimi Paschal, M.S., San Joaquin Valley College; Elizabeth Pearsall, Ph.D., CPhT, Virginia College, Biloxi, MS; Michael Perun, B.S., Lincoln Technical Institute, Allentown, PA; and Lora Plank, M.S.N., R.N., CNOR, CST, CPhT, Ivy Tech Community College, Valparaiso, IN.

Reviewers of the previous editions, whose insights and feedback also helped shape the sixth edition, are James Austin, R.N., B.S.N., CPhT, Program Chair, Pharmacy Technology Program, Weatherford College; Ashley Bures, CVS; CVS Pharmacy; Marisa Fetzer, Institute of Technology; Claudia Johnson, M.S.N., R.N.C., A.P.N., Allied Health Instructor, Pharmacy Technician Program, Polytech Adult Education; Mary Anna Marshall, CPhT, Instructor for PTCB Examination Review, CVS/Pharmacy Regional Intern Coordinator, Pharmacy Technician Program, Hanover High School; Russell C. McGuire, Ph.D., Vice President of Allied Health Programs, Education Corporation of America; Tony David Ornelas, CPhT; Alexandria Zarrina Ostowari, Course Developer, Allied Business School; Becky Schonscheck, Director of Curriculum Development, Anthem Education Group; Jacqueline T. Smith R.N., CPhT, Department Chair–Pharmacy Technician, National College; Peter E. Vonderau, Sr., R.Ph., Pharmacy Manager, Scolari's Food & Drug, Inc.; Walgreen Co.; and Deborah Zenzal , R.N. M.S. CPC CCS-P RMA, Department Chair, Allied Health, Penn Foster Schools.

Finally, we would like to thank Brenda Carmichael for designing and preparing the cover; Elizabeth Budd for preparing the index; Rayna Bailey, Will Kelley, Marta Martins, Adam Jones, David Ferguson, and Chrissy Morton of Morton Publishing for their help with the sixth edition; and Doug Morton, whose sponsorship makes this book possible.

Perspective Press

Dear Student or Instructor,

The American Pharmacists Association (APhA), the national professional society of pharmacists in the United States, and Morton Publishing Company, a publisher of educational texts and training materials in health care, are pleased to present this outstanding workbook, *The Pharmacy Technician Workbook and Certification Review*, Sixth Edition. It is one of a series of distinctive texts and training materials for basic pharmacy and pharmacology training that is published under this banner: *American Pharmacists Association Basic Pharmacy and Pharmacology Series.*

Each book in the series is oriented toward developing an understanding of fundamental concepts. In addition, each text presents applied and practical information on the skills necessary to function effectively in positions such as technicians and medical assistants who work with medications and whose role in health care is increasingly important. Each of the books in the series uses a visual design to enhance understanding and ease of use. We think you will find them valuable training tools.

The American Pharmacists Association and Morton Publishing thank you for using this book and invite you to look at other titles in this series, which are listed below.

Thomas E. Menighan, BSPharm, MBA
Executive Vice President
Chief Executive Officer
American Pharmacists Association

David M. Ferguson
President
Morton Publishing Company

TITLES IN THIS SERIES:

The Pharmacy Technician, Sixth Edition
The Pharmacy Technician Workbook and Certification Review, Sixth Edition
Medication Workbook for Pharmacy Technicians: A Pharmacology Primer

The Pharmacy Technician

Workbook and Certification Review

— 1 —

PHARMACY & HEALTH CARE

KEY CONCEPTS

Test your knowledge by covering the information in the right-hand column.

pharmacology The study of drugs, their properties, uses, application, and effects (from the Greek *pharmakon*: drug, and *logos*: word or thought).

herbal medicine People have used drugs derived from plants to treat illnesses and other physical conditions for thousands of years. The ancient Greeks used the bark of a white willow tree to relieve pain. The bark contained salicylic acid, the natural forerunner of the active ingredient in aspirin.

quinine The first useful drug in the treatment of malaria, one of mankind's most deadly diseases. It was extracted from the bark of a Peruvian tree, the cinchona.

cocaine The first effective local anesthetic.

digitalis The drug of the foxglove plant, which has been widely used in treating heart disease.

synthetic drugs Drugs created by reformulating simpler chemicals into more complex ones, creating a new chemical not found in nature.

average life span By 2017, this measure of health is expected to have increased by almost 40 years since 1900.

germ theory The theory that microorganisms cause food spoilage.

insulin The hormone that lowers blood sugar in the treatment of diabetes—one of the great discoveries in medicine in the twentieth century.

penicillin The first antibiotic.

polio vaccine The use of an injectable vaccine made from killed polio virus and an oral polio vaccine made from a weakened form of live polio virus was important to prevent the spread of this crippling and often fatal disease.

Human Genome Project An attempt to map the entire DNA sequence in the human genome. This information will provide a better understanding of hereditary diseases and how to treat them.

pharmacist education and training To become a pharmacist in the United States, an individual must have earned a Doctor of Pharmacy degree from an accredited college of pharmacy (of which there are about 130 in the United States), pass a state licensing exam (in some states), and perform experiential training under a licensed pharmacist. Once licensed, the pharmacist must receive continuing education to maintain their license.

cost control A significant trend in recent health care has been the effort to control the cost of prescription drugs, one aspect of which is the use of closed "formularies" that rely substantially on substituting generic drugs in place of more expensive brands.

computerization Pharmacy computer systems put customer profiles, product, inventory, pricing, and other essential information within easy access. One result has been that pharmacies and pharmacists dispense more prescriptions and information than ever before.

STUDY NOTES

Use this area to write important points you'd like to remember.

FILL IN THE KEY TERM

Use these key terms to fill in the correct blank. Answers are at the end of the book.

antibiotic
antitoxin
cocaine
digitalis
hormones
human genome
long-term care
Medicare Modernization Act
MTM services
penicillin
pharmacognosy
pharmacology
quinine
salicylic acid
synthetic

1. cocaine: A local anesthetic found in coca leaves.
2. Salicylic acid: The natural drug derived from the bark of a white willow tree, used by the ancient Greeks to relieve pain, and the natural forerunner to the active ingredient in aspirin.
3. quinine: A drug for malaria found in cinchona bark.
4. human genome: The complete set of genetic information in the human cell.
5. Long-term care: Residence facilities that provide care on a long-term rather than short-term basis.
6. pharmacology: The study of drugs—their properties, uses, application, and effects.
7. penicillin: A drug produced by a fungus that kills bacteria.
8. MTM: Services provided by a pharmacist that look at all the medications a patient is taking.
9. antitoxin: A substance that acts against a toxin in the body.
10. antibiotic: A substance that harms or kills microorganisms like bacteria and fungi.
11. Hormone: Chemicals produced by the body that regulate body functions and processes.
12. pharmacogny: The study of physical, chemical, biochemical, and biological properties of drugs.
13. Medicare: Expanded the role of the pharmacist to provide MTM services to some Medicare patients.
14. digitalis: The drug of the foxglove plant, which has been widely used in treating heart disease.
15. Synthetic: Combining simpler chemicals into more complex compounds not found in nature.

TRUE/FALSE

Indicate whether the statement is true or false in the blank. Answers are at the end of the book.

F 1. The natural drug that is the forerunner to aspirin comes from the coca leaves.

T 2. Most drugs used today are made synthetically.

T 3. Digitalis comes from the foxglove plant.

F 4. Cocaine was the first synthetic hormone.

T 5. In 1900 the average American lived only until their 40s.

T 6. More pharmacists and technicians are employed in community pharmacies than in any other setting.

F 7. The second largest area of employment for pharmacists and technicians is long-term care.

F 8. The field of biotechnology has become the least dynamic area of pharmaceutical research and development.

T 9. According to Gallup Polls, pharmacists consistently rank as one of the most highly trusted and ethical professions in the United States.

F 10. Smartphones are not currently used to access drug information.

EXPLAIN WHY

Explain why these statements are true or important. Check your answers in the text. Discuss any questions you may have with your instructor.

1. Give at least three reasons why synthetic drugs are important.
2. Why was the use of anesthesia revolutionary?
3. Why was Paracelsus's work important?
4. Why was penicillin a major benefit in wartime?
5. Why is the Human Genome Project important to pharmacology?
6. Why are drug patents important?
7. Why are pharmacists among the most trusted professionals?

CHOOSE THE BEST ANSWER

Answers are at the end of the book.

1. The drug digitalis comes from the foxglove plant and is used to treat some ________________ conditions.
   a. liver
   b. kidney
   c. heart
   d. lung

2. The first publicized operation using general anesthesia was performed using __________ as the anesthetic.
   a. cocaine
   b. foxglove
   c. quinine
   d. ether

3. The field of ________________ has resulted from the study of the human genome.
   a. pharmacology
   b. biotechnology
   c. natural medicine
   d. discovery

4. The drug form of coca leaves is used for
   a. local anesthesia.
   b. diabetes.
   c. heart disease.
   d. hypertension.

5. Dioscorides was a/an __________ physician.
   a. Egyptian
   b. Persian
   c. Greek
   d. Spanish

6. Carl Koller discovered cocaine was useful for local anesthesia in ________surgery.
   a. dental
   b. dermatologic
   c. gynecologic
   d. eye

7. ________________ developed an oral polio vaccine.
   a. Watson and Crick
   b. Sabin
   c. Fleming
   d. Hippocrates

8. An authoritative listing of drugs and issues related to their use is a (an)
   a. pharmacopeia.
   b. materia medica.
   c. panacea.
   d. Sumerian.

9. Care that is managed by an insurer is
   a. home care.
   b. managed care.
   c. short-term care.
   d. long-term care.

10. Pharmacists are paid approximately ________________ for MTM services.
    a. $10–$20 per hour
    b. $30–$40 per hour
    c. $1–$3 per minute
    d. $5–$10 per minute

STUDY NOTES

Use this area to write important points you'd like to remember.

— 2 —
THE PROFESSIONAL PHARMACY TECHNICIAN

KEY CONCEPTS

Test your knowledge by covering the information in the right-hand column.

job responsibilities Pharmacy technicians perform essential tasks that do not require the pharmacist's skill or expertise. Specific responsibilities and tasks differ by setting and are described in writing by each employer through job descriptions, policy and procedure manuals, and other documents.

supervision Pharmacy technicians work under the direct supervision of a licensed pharmacist who is legally responsible for their performance.

pharmacist counseling Having technicians assist the pharmacist frees the pharmacist for activities that require a greater level of expertise, such as counseling with patients.

scope of practice What individuals may and may not do in their jobs is often referred to as their "scope of practice."

employment opportunities Like pharmacists, most pharmacy technicians are employed in community pharmacies and hospitals. However, they are also employed in clinics, home care, long-term care, mail order prescription pharmacies, and various other settings.

specialized job In various hospital and other environments, there are specialized technician jobs, which require more advanced skills developed from additional education, training, and experience.

trustworthiness Pharmacy technicians are entrusted with confidential patient information, dangerous substances, and perishable products.

errors Drugs, whether prescription or over the counter, can be dangerous if misused, and mistakes by pharmacy technicians can be life threatening.

Health Insurance Portability and Accountability Act (HIPAA) Pharmacy technicians are legally responsible for the privacy and security of protected health information (PHI).

math skills Pharmacy technicians routinely perform mathematical calculations in filling prescriptions and other activities.

terminology Pharmacy technicians must learn the specific pharmaceutical terminology that will be used on the job.

teamwork Pharmacy technicians must be able to communicate, cooperate, and work effectively with others.

standards There is no federal standard for pharmacy technician training or competency. However there are state and employer standards that must be met.

certification A valuable career step for pharmacy technicians is getting certification by an appropriate organization or body. It verifies an individual's competence as a technician, and indicates a high level of knowledge and skill. In the United States, the Pharmacy Technician Certification Exam (PTCE) and the Exam for the Certification of Pharmacy Technicians (ExCPT) are national exams that lead to technician certification.

communication Good communication is an important part of teamwork. Taking time to communicate relevant information in the pharmacy is important.

preventing errors Preventing errors is an important goal for every pharmacy. Pharmacy technicians play an important role in preventing errors.

STUDY NOTES

Use this area to write important points you'd like to remember.

FILL IN THE KEY TERM

Use these key terms to fill in the correct blank. Answers are at the end of the book.

ASHP
certification
continuing education
cultural competence
ExCPT
HIPAA
patient welfare
performance review
personal inventory
pharmacist
PTCE
scope of practice
tall man lettering
technicians

1. ____________________ : What individuals may and may not do in their jobs.
2. ____________________ : To assess characteristics, skills, qualities, etc.
3. ____________________ : A term that describes individuals who are able to provide care to patients of diverse backgrounds.
4. ____________________ : The most important consideration in health care.
5. ____________________ : Exam offered by the Pharmacy Technician Certification Board.
6. ____________________ : A legal proof or document that an individual meets certain objective standards, usually provided by a neutral professional organization.
7. ____________________ : Individuals who are given a basic level of training designed to help them perform specific tasks.
8. ____________________ : Capital letters to emphasize the different parts of drug names.
9. ____________________ : Technicians always work under their direct supervision.
10. ____________________ : Employer documents employee's job competency.
11. ____________________ : Has model curriculum for technician training.
12. ____________________ : Exam given by the National Healthcareer Association.
13. ____________________ : A critical element in maintaining competency for pharmacy technicians.
14. ____________________ : Law that makes health-care providers responsible for the privacy and security of a patient's health information.

TRUE/FALSE

Indicate whether the statement is true or false in the blank. Answers are at the end of the book.

_____ 1. Specific technician responsibilities differ by setting and job description.

_____ 2. Technicians may sometimes provide counseling services to patients.

_____ 3. Most pharmacy technicians are employed in mail order pharmacies.

_____ 4. Mathematics skills are very important to the pharmacy technician.

_____ 5. It is essential for technicians to have good interpersonal skills.

_____ 6. The U.S. government sets standards for technician training.

_____ 7. PHI is considered public information.

_____ 8. As a technician, your employer is legally responsible for your performance.

_____ 9. Good communication is an important part of teamwork.

_____ 10. TJC recommends using the abbreviation U for unit.

EXPLAIN WHY

Explain why these statements are true or important. Check your answers in the text. Discuss any questions you may have with your instructor.

1. Give at least two reasons technicians must work under the supervision of a pharmacist.
2. Why is knowing your "scope of practice" important?
3. Why is dependability important?
4. Why is it important that pharmacy technicians understand their responsibilities under the 1996 Health Insurance Portability and Accountability Act (HIPAA)?
5. Why should technicians have math skills?
6. Why are interpersonal skills important?
7. Why is certification a good idea for technicians?
8. Why is continuing education valuable for technicians?

IN THE WORKPLACE

These sample job descriptions may help you in your career choice.

Sample Pharmacy Technician Job Description – Community Pharmacy

GENERAL INFORMATION

Under the direct supervision of the pharmacist, THE PHARMACY TECHNICIAN assists the pharmacist with day-to-day activities

ROLE AND RESPONSIBILITIES

- Answering telephones and screening telephone calls for the pharmacist
- Assisting patients who are dropping off or picking up prescriptions
- Assisting the pharmacist with filling and labeling prescriptions
- Communicating with third-party plans for purposes of processing prescription claims
- Compounding oral liquid medications, ointments, and creams
- Contacting prescribers and their agents to obtain refill authorizations
- Creating and maintaining patient profiles
- Entering data into the computer
- Ordering stock and placing orders on shelves
- Preparing the pharmacy for inventories
- Processing and verifying paperwork
- Repackaging bulk medications

QUALIFICATIONS AND EDUCATION REQUIREMENTS

- Ability to perform basic pharmacy calculations with accuracy
- Ability to respect and maintain confidentiality of patient information
- Ability to type at least 35 words per minute
- Knowledge of brand names and generic names of medications
- Knowledge of third-party payment systems for prescriptions and pharmacist provide services
- Professionalism
- Strong communication skills
- Understanding of medical terminology
- High school diploma or GED
  National certification must be obtained within 6 months of hire

Sample Pharmacy Technician Job Description – Hospital Pharmacy

GENERAL INFORMATION

Under the direct supervision of a pharmacist, and in compliance with federal and state laws and department policies and procedures, the pharmacy technician performs pharmacy-related functions.

RESPONSIBILITIES

- Answer telephones and screen telephone calls
- Assist other pharmacy technicians as needed
- Collect quality assurance data
- Compound large volume parenterals
- Compound ointments, creams, and oral medications
- Compound total parenteral nutrition solutions
- Enter medication orders into the pharmacy computer system
- Fill patient medication cassettes
- Order drugs and supplies from the store room
- Package and prepare drugs for investigational use
- Perform monthly nursing unit inspections and maintain associated records
- Pick up unused medications from nursing units
- Prepare chemotherapy drugs
- Prepare prepackaged bulk medications
- Prepare prescriptions for outpatient use
- Receive drug orders and stock shelves
- Rotate through all work areas of the pharmacy
- Transport medications, drug delivery devices, and other pharmacy materials, between the pharmacy and the nursing units and clinics within the hospital

QUALIFICATIONS AND EDUCATION REQUIREMENTS

- Ability to work as a team member
- Ability to perform pharmacy calculations accurately
- Ability to type 35 words per minute accurately
- Attention to detail
- Good communication skills
- High school diploma or GED
- Knowledge of basic pharmacy practices and procedures
- Knowledge of medications and medical supplies
- Knowledge of recordkeeping techniques
- National certification
- Valid state pharmacy technician license (required in some states)

EXPERIENCE

- Must have one year of pharmacy experience, preferably in a hospital

IN THE WORKPLACE

This sample resume may help you when applying for a position as a pharmacy technician in a community pharmacy.

Ashley Stevens
4444 Main St.
Tucson, AZ 84710
480-345-6789
astevens@asdf.com

Objective

Pharmacy technician position in a community pharmacy

Experience

Receptionist, Law Office of John Smith — May 2015 - Present

- Received employee of the month award

Sales Associate, Sally's Flowers — September 2011 – May 2015

Volunteer, City Hospital — June 2011 – September 2011

- Tucson, AZ
- Assisted at visitor information desk

Education

University of Arizona, Tucson, AZ — August 2011 – May 2015

- Bachelor of Science (Chemistry)
- Cum laude

Central High School, Phoenix, AZ — August 2007 – June 2011

References

References are available on request.

IN THE WORKPLACE

This sample resume may help you when applying for a position as a pharmacy technician.

PENNY J. STEVENS

2543 Gaylord St.| Jacksonville, FL | 904-445-5554 | penny.stevens@asdfgh.com

EDUCATION

University of North Florida — September 2014 - Present
Jacksonville, FL
Completed 63 Semester Hours

City High School — September 2000 – May 2014
Jacksonville, Florida

EXPERIENCE

Pharmacy Technician — July 2014 - Present
Grey's Pharmacy

- Assist pharmacist with dispensing 150 prescriptions daily; place orders, stock shelves, provide excellent customer service

Server — May 2014 – July 2014
Bob's Restaurant

- Took orders and served food to patrons

Server — May 2012 – May 2014
Joe's Fast Food

- Took orders, filled orders, ran cash register

COMPUTER SKILLS

Microsoft Windows, Word, Excel

REFERENCES

Available on request

IN THE WORKPLACE

This sample resume may help you when applying for a position as a pharmacy technician in a hospital pharmacy.

STEVEN A. CARTER
2511 West Rd., Iowa City, IA 52240
319-555-1111
sacarter@abcdef.com

OBJECTIVE

Pharmacy technician position in a hospital pharmacy

EDUCATION

University of Iowa, Iowa City, IA
Pre-Pharmacy 2014-present

University of Iowa, Iowa City, IA
B.A. in English 2014

WORK EXPERIENCE

Pharmacy Technician
Smith Pharmacy, Des Moines, IA June 2014 - present

Lawn mowing June 2012- June 2014
Charlie's Lawn Care (summers)

CERTIFICATION

Certified Pharmacy Technician
Pharmacy Technician Certification Board (PTCB) Awarded December 2014

REFERENCES

References available upon request

IN THE WORKPLACE

This sample resume may help you when applying for admission to pharmacy school.

Jason N. James

1225 White Rd.
Oakland, CA 94611
510-345-6789

Objective

To gain admission to the University of California–San Diego School of Pharmacy

Experience

June 2012 – present	Pharmacy Technician	Harbor Pharmacy, Oakland, CA
January 2012 – May 2012	Pharmacy Technician	Bayview Hospital, Oakland, CA
June 2011 – January 2012	Hospital Transportation	Bayview Hospital, Oakland, CA
June 2008 – June 2011	Waiter/Server	Bob's Burgers, Oakland, CA

Education

September 2004 – June 2008	Diploma	Smith High School, Oakland, CA

Certification

2012 – present	Certified Pharmacy Technician	Pharmacy Technician Certification Board (PTCB)

Community Service

June 2012 – present	Volunteer Coach	YMCA

References

References are available on request.

IN THE WORKPLACE ACTIVITIES

1. Use Microsoft Word to create a letter to apply for a job.

2. Prepare a resume for an entry level job as pharmacy technician.

3. Use the following form to critique your classmates on the suitability of their appearance for the work environment. Choose one of the following pharmacy technician workplaces: chain community pharmacy, hospital pharmacy, independent community pharmacy, home infusion pharmacy, other pharmacy related environment. Use the following form to critique your classmates on the suitability of their appearance for the work environment.

Appropriate clothing
Appropriate shoes
Appropriate hair
Other comments:

4. Conduct the following tasks in a predetermined amount of time. Use a stopwatch or smart phone app for timing purposes.

Pour 6 ounces in an amber bottle
Pur 4 ounces in an amber bottle
Use a counting tray and spatula to count 30 capsules
Use a counting tray and a spatula to count 30 tablets

IN THE WORKPLACE ACTIVITIES (CONT'D)

5. Use the following form to interview a patient and obtain the necessary information so the patient's profile can be added to the pharmacy's computer system.

Pharmacy Patient Profile Form

Last Name	First Name	Middle Initial
Physical Address	City	State
E-mail Address	Phone Number (Including Area Code)	Zip Code

Insurance Information

Member ID Number	Group Number
Policy Holder Name	Policy Holder Date of Birth

Health Profile

Please provide information below for each family member. If additional space is needed, please check here ☐ and attach a separate sheet with the additional information.

	CARDHOLDER	SPOUSE	DEPENDENT	DEPENDENT	DEPENDENT
LAST NAME					
FIRST NAME					
MIDDLE INITIAL					
DATE OF BIRTH (MM/DD/YYYY)					
SEX					

Drug Allergies: please check the appropriate box(es) where a drug allergy is known

Please provide information below for each family member. If additional space is needed, please check here ☐ and attach a separate sheet with the additional information.

	CARDHOLDER	SPOUSE	DEPENDENT	DEPENDENT	DEPENDENT
PENICILLIN	☐	☐	☐	☐	☐
CODEINE	☐	☐	☐	☐	☐
ASPIRIN	☐	☐	☐	☐	☐
SULFA	☐	☐	☐	☐	☐
OTHER					
NO KNOWN ALLERGIES	☐	☐	☐	☐	☐

Disease States: please check the appropriate box(es) for known medical conditions

Please provide information below for each family member. If additional space is needed, please check here ☐ and attach a separate sheet with the additional information.

	CARDHOLDER	SPOUSE	DEPENDENT	DEPENDENT	DEPENDENT
HIGH BLOOD PRESSURE	☐	☐	☐	☐	☐
DIABETES	☐	☐	☐	☐	☐
THYROID	☐	☐	☐	☐	☐
ASTHMA	☐	☐	☐	☐	☐
SEIZURES	☐	☐	☐	☐	☐
GLAUCOMA	☐	☐	☐	☐	☐
OTHER					
NO KNOWN DISEASES	☐	☐	☐	☐	☐

Current Medications: please list known current medications

Please provide information below for each family member. If additional space is needed, please check here ☐ and attach a separate sheet with the additional information.

	CARDHOLDER	SPOUSE	DEPENDENT	DEPENDENT	DEPENDENT

IN THE WORKPLACE ACTIVITIES (CONT'D)

6. A flu clinic is scheduled at your pharmacy for this Friday. An order will also be delivered. And it is the first of the month, so the pharmacy will be very busy with prescriptions. There are five technicians that work at the pharmacy. If the pharmacy is open seven days a week, discuss some strategies to get through this unusually busy day on Friday.

7. Work in groups of three for the following role-play exercises. One person is the patient. One person is a pharmacy technician. One person is an observer. The observer provides a critique after watching the technician convey information in the scenario to the patient.
   a. Simulate a telephone conversation in which the patient is calling in a refill.
   b. The patient comes to pick up his or her prescription. The medication is out of stock.
   c. The patient comes to pick up his or her prescription. Their prescription is out of refills. The doctor's office will not authorize refills because the patient needs to make an appointment.
   d. The patient comes to pick up his or her prescription. The prescription is filled, but the price is $50 more than the patient was expecting to pay. The insurance covered the prescription, but the prescription has a higher co-pay than the patient expected.
   e. The patient refuses to provide his/her address and phone number. The pharmacy technician attempts to get all the necessary demographic information from the patient.
   f. The patient dropped off his or her prescription, and wants to know if it is ready. The medication is not commercially available so it needs to be compounded by the pharmacist. Because one ingredient must be ordered, the medication will not be ready for two days.
   g. The patient is hearing-impaired and brings a prescription to the pharmacy to be filled.
   h. The patient is visually impaired brings and brings a prescription to the pharmacy to be filled.

CHOOSE THE BEST ANSWER

Answers are at the end of the book.

1. When technicians perform appropriate essential tasks, this allows the pharmacist time for tasks requiring more advanced professional expertise such as
   a. telephoning insurance companies.
   b. consulting with patients.
   c. counting tablets.
   d. ringing the cash register.

2. Taking routine patient information is a duty of the
   a. consultant.
   b. cashier.
   c. pharmacist.
   d. pharmacy technician.

3. Pharmacy technicians should be detail-oriented. This means
   a. technicians' work can always be done by pharmacists.
   b. technicians are not as important as pharmacists.
   c. patients must receive medications exactly as they are prescribed.
   d. tardiness is acceptable.

4. Do pharmacy technicians need to maintain good physical and mental health?
   a. Yes, to decrease the chance of making serious mistakes.
   b. Yes, because some belong to labor unions.
   c. No. Pharmacists are always responsible for the technician, so technicians don't have to worry about getting enough sleep.
   d. No. Pharmacists have more education and so will catch all errors made.

5. According to HIPAA, it is okay to
   a. discuss patient information within ear-shot of other patients.
   b. bill insurance using HIPAA-compliant EDI.
   c. bill insurance using nonsecure EDI.
   d. casually discuss a patient's condition with a patient's spouse.

6. ______________ refers to being qualified for or capable of performing a task or job.
   a. Scope of practice
   b. Compounding
   c. Dependability
   d. Competent

7. ______________________ are regularly scheduled events to monitor and document technician competency.
   a. PTCE
   b. PTCB
   c. Performance reviews
   d. TJC

8. __________________ is a legal proof or document that an individual meets certain objective standards, usually provided by a neutral professional organization.
   a. Registration
   b. Certification
   c. Documentation
   d. Prior authorization

9. A/an example of complementary and alternative medicine are
   a. NSAIDs.
   b. ACE inhibitors.
   c. dietary supplements
   d. phenothiazines.

10. An example of a brand-name extension is
    a. Valium®.
    b. Xanax®.
    c. Tylenol®
    d. furosemide.

— 3 —

PHARMACY LAWS, REGULATIONS, & ETHICS

KEY CONCEPTS

Test your knowledge by covering the information in the right-hand column.

Food and Drug Administration	The leading enforcement agency at the federal level for regulations concerning drug products.
Drug Enforcement Administration	The agency that controls the distribution of drugs that may be easily abused.
Food and Drug Act of 1906	Prohibited interstate commerce in adulterated or misbranded food, drinks, and drugs. Government preapproval of drugs is required.
1938 Food, Drug and Cosmetic (FDC) Act	In response to the fatal poisoning of 107 people, primarily children, by an untested sulfanilamide concoction, this comprehensive law requires new drugs be shown to be safe before marketing.
1951 Durham-Humphrey Amendment	This law defines what drugs require a prescription by a licensed practitioner and requires them to include this legend on the label: "Caution: Federal Law prohibits dispensing without a prescription."
1962 Kefauver-Harris Amendments	Requires drug manufacturers to provide proof of both safety and effectiveness before marketing the drug.
1970 Poison Prevention Packaging Act	Requires childproof packaging on all controlled and most prescription drugs dispensed by pharmacies.
1970 Controlled Substances Act (CSA)	The CSA classifies drugs that may be easily abused and restricts their distribution. It is enforced by the Drug Enforcement Administration (DEA) within the Justice Department.
1990 Omnibus Budget Reconciliation Act (OBRA)	Among other things, this act required pharmacists to offer counseling to Medicaid patients regarding medications, effectively putting the common practice into law.

1996 Health Insurance Portability and Accountability Act (HIPAA) Provided broad and stringent regulations to protect patients' privacy.

placebos Inactive substances, not real medications, that are used to test the effectiveness of drugs.

new drugs All new drugs, whether made domestically or imported, require FDA approval before they can be marketed in the United States.

clinical tests Tests on proposed new drugs (investigational drugs) are "controlled" by comparing the effect of a proposed drug on one group of patients with the effect of a different treatment on other patients.

blind tests Patients in a trial are always "blind" to the treatment, i.e., they are not told which control group they are in. In a "double-blind" test, neither the patients nor the physicians know what the medication is.

patent protection A patent for a new drug gives its manufacturer an exclusive right to market the drug for a specific period of time under a brand name. A drug patent is in effect for 17 years from the date of the drug's discovery. The Hatch-Waxman Act of 1984 provided for up to five-year extensions of patent protection to the patent holders to make up for time lost while products went through the FDA approval process.

generics Once a patent for a brand drug expires, other manufacturers may copy the drug and release it under its pharmaceutical, or "generic," name.

product labeling In addition to a container label, manufacturers' prescription drugs must be accompanied by product labeling that can include package inserts and MedGuides.

prescription drug labels The minimum requirements on prescription labels for most drugs are as follows: name and address of dispenser, prescription serial number, date of prescription or filling, name of prescriber, name of patient, directions for use, and cautionary statements.

NDC (National Drug Code) number The number assigned by the manufacturer. Each NDC number has three parts, or sets, of numbers: The first set indicates the manufacturer; the next set indicates the medication, its strength, and dosage form; the last set indicates the package size.

controlled substance A drug that has the potential to be abused and for which distribution is controlled by one of five "schedules."

control classifications Manufacturers must clearly label controlled drugs with their control classification.

KEY CONCEPTS

Test your knowledge by covering the information in the right-hand column.

DEA number/formula	The **number** all prescribers of controlled substances are assigned and that must be used on all controlled drug prescriptions. It has two letters followed by seven single-digit numbers, e.g., AB1234563. The **formula** for checking a DEA number on a prescription form is: if the sum of the first, third and fifth digits is added to twice the sum of the second, fourth, and sixth digits, the total should be a number whose last digit is the same as the last digit of the DEA number.
risks of approved drugs	There is always the risk that an approved drug may produce adverse side effects when used on a larger population.
recalls	Recalls are, with a few exceptions, voluntary on the part of the manufacturer. There are three classes of recalls: I. where there is a strong likelihood that the product will cause serious adverse effects or death; II. where a product may cause temporary but reversible adverse effects, or in which there is little likelihood of serious adverse effects; III. where a product is not likely to cause adverse effects.
MedWatch	FDA reporting program for health-care professionals to report adverse effects that occur from the use of an approved drug or other medical product. The MedWatch Online Voluntary Reporting Form 3500 is used for FDA-regulated drugs, biologics, medical devices, and special nutritional products and cosmetics.
criminal law	Laws pertaining to a wrong to society.
civil law	Laws pertaining to a wrong to an individual.
liability	Legal liability means you can be prosecuted for misconduct.
negligence	Failing to do something that should or must be done.

CONTROLLED SUBSTANCE SCHEDULES

The five control schedules are as follows:*

Schedule I:

- Each drug has a high potential for abuse and no accepted medical use in the United States. It may not be prescribed. Heroin, various opium derivatives, and hallucinogenic substances are included on this schedule.

Schedule II:

- Each drug has a high potential for abuse and may lead to physical or psychological dependence, but also has a currently accepted medical use in the United States. Amphetamines, opium, cocaine, methadone, and various opiates are included on this schedule.

Schedule III:

- Each drug's potential for abuse is less than those in Schedules I and II and there is a currently accepted medical use in the United States, but abuse may lead to moderate or low physical dependence or high psychological dependence. Anabolic steroids and various compounds containing limited quantities of narcotic substances such as codeine are included on this schedule.

Schedule IV:

- Each drug has a low potential for abuse relative to Schedule III drugs and there is a currently accepted medical use in the United States, but abuse may lead to limited physical dependence or psychological dependence. Phenobarbital, the sedative chloral hydrate, and the anesthetic methohexital are included in this group.

Schedule V:

- Each drug has a low potential for abuse relative to Schedule IV drugs and there is a currently accepted medical use in the United States, but abuse may lead to limited physical dependence or psychological dependence. Compounds containing limited amounts of a narcotic such as codeine are included in this group.

**21 USC Sec. 812. Source: http://www.deadiversion.usdoj.gov/21cfr/21usc/812.htm. Note: These schedules are revised periodically. It is important to refer to the most current schedule.*

Controlled-Substance Prescriptions

Controlled-substance prescriptions have greater requirements at both federal and state levels than other prescriptions, particularly Schedule II drugs. On controlled-substance prescriptions, the DEA number must appear on the form and the patient's full street address must be entered.

On Schedule II prescriptions, the form must be signed by the prescriber. In many states, there are specific time limits that require Schedule II prescriptions be promptly filled. Generally, quantities are limited and no refills are allowed.

Federal requirements for Schedules III–V are less stringent than for Schedule II. *For example, Schedules III–V prescriptions may be refilled up to five times within six months.* However, state and other regulations may be stricter than federal requirements, so it is necessary to know the requirements for your specific job setting.

APPROVED DRUG PRODUCTS WITH THERAPEUTIC EQUIVALENCE EVALUATIONS (THE "ORANGE BOOK")

The FDA annually publishes *Approved Drug Products With Therapeutic Equivalence Evaluations* (the "Orange Book"), and updates it throughout the year online at http://www.accessdata.fda.gov/scripts/cder/ob/default.cfm. The "Orange Book" provides a two-letter evaluation code to allow users to determine whether the FDA has evaluated a particular product as therapeutically equivalent to other pharmaceutically equivalent products (first letter) and to provide additional information on the basis of FDA's evaluations (second letter).

"**A**" drug products are those the FDA considers to be therapeutically equivalent to other pharmaceutically equivalent products. There are (1) no known or suspected bioequivalence problems or (2) the actual or potential bioequivalence problems have been resolved with adequate in vivo and/or in vitro evidence supporting bioequivalence.

"**B**" drug products are those the FDA considers not to be therapeutically equivalent to other pharmaceutically equivalent products. Often, the problem is with specific dosage forms rather than with the active ingredient.

A listing of the codes is below:

AA Products in conventional dosage forms not presenting bioequivalence problems
AB Products meeting bioequivalence requirements
AN Solutions and powders for aerosolization
AO Injectable oil solutions
AP Injectable aqueous solutions
AT Topical products

BC Extended-release tablets, extended-release capsules, and extended-release injectables
BD Active ingredients and dosage forms with documented bioequivalence problems
BE Delayed-release oral dosage forms
BN Products in aerosol-nebulizer drug delivery systems
BP Active ingredients and dosage forms with potential bioequivalence problems
BR Suppositories or enemas for systemic use
BS Products with drug standard deficiencies
BT Topical products with bioequivalence issues
BX Insufficient data

There are some products that cannot be described with just the two letter code, i.e., a more complete explanation is needed. These products often have problems with identity, analytical methodology, or bioequivalence standards. For some of these products, bioequivalence has not been established or no generic product is currently available.

FILL IN THE KEY TERM

Use these key terms to fill in the correct blank. Answers are at the end of the book.

adverse effect	**legend drug**	**OTC drugs**
autonomy	**liability**	**pediatric**
beneficence	**NDC (national drug ode)**	**placebo**
Controlled Substances Act	**negligence**	**recall**
DEA number	**Protected Health Information (PHI)**	

1. __DEA__: The number all prescribers of controlled substances are assigned and that must be used on all controlled drug prescriptions.
2. __Benebicence__: The actions of the health-care provider should be in the best interest of the patient.
3. __place bo__: An inactive substance given in place of a medication.
4. __adverse ebbect__: An unintended side affect of a medication that is negative or in some way injurious to a patient's health.
5. __legend drug__: Any drug that requires a prescription and this "legend" on the label: Rx only.
6. __negugence__: Failing to do something you should have done.
7. __pediatric__: Having to do with the treatment of children.
8. __autonomy__: Patients have the right to choose their treatment.
9. __PHI__: Any personal information that could be used to identify an individual or their health history.
10. __liabicity__: Means you can be prosecuted for misconduct.
11. __recall__: The action taken to remove a drug from the market and have it returned to the manufacturer.
12. __controlled Sub Act__: 1970 law that established schedules of controlled substances.
13. __NDC__: The number on a manufacturer's label indicating the manufacturer and product information.
14. __OTC drugs__: Drugs that do not require a prescription.

TRUE/FALSE

Indicate whether the statement is true or false in the blank. Answers are at the end of the book.

____ 1. Childproof packaging was required by the Drug Enforcement Administration.

____ 2. The CMEA sets daily and monthly restrictions on the sale of pseudoephedrine.

____ 3. Only about 25% of drugs tested on humans are approved for use by the FDA.

____ 4. Over-the-counter medications do not require a prescription but sometimes prescriptions are written for them.

____ 5. The prescriber's name must appear on the label of a dispensed prescription container.

____ 6. A Class III recall is most likely to cause harm or death.

____ 7. HITECH effectively amends HIPAA.

____ 8. Schedule III, IV, and V drugs may be stored openly on shelves in retail and hospital settings.

____ 9. All controlled substances must be ordered using a DEA controlled substance order form.

____ 10. Justice means fairness and equality should be applied when providing care to all patients.

EXPLAIN WHY

Explain why these statements are true or important. Check your answers in the text. Discuss any questions you may have with your instructor.

1. Why is blind testing used in the drug approval process?
2. Give three reasons why OTC labels should be clear and understandable.
3. Why are some drugs "controlled" by the DEA?
4. Why are some drug patents extended past the original 17-year period?
5. Why would a manufacturer want to recall a drug product?
6. Give three reasons why failing to do something could result in a criminal charge of negligence.

MATCH THE TERMS — CONTROLLED SUBSTANCES AND RECALLS

Use these key terms to fill in the correct blank. Answers are at the end of the book.

Schedule I Drugs
Schedule II Drugs
Schedule III Drugs
Schedule IV Drugs
Schedule V Drugs
Class I Recall
Class II Recall
Class III Recall

1. ________________: Amphetamines, opium, cocaine, methadone, and various opiates are included on this schedule.

2. ________________: Anabolic steroids and various compounds containing limited quantities of narcotic substances such as codeine are included on this schedule.

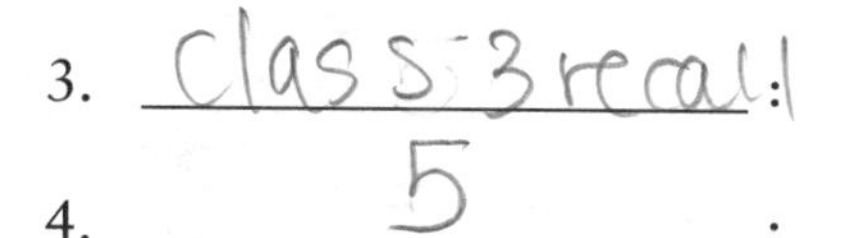

3. ________________: When a product is not likely to cause adverse effects.

4. ________________: Compounds containing limited amounts of a narcotic such as codeine are included in this group.

5. ________________: When a product may cause temporary but reversible adverse effects, or in which there is little likelihood of serious adverse effects.

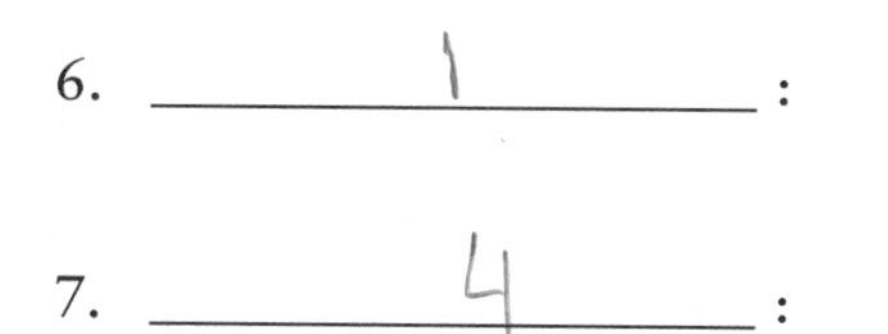

6. ________________: Heroin, various opium derivatives, and hallucinogenic substances are included on this schedule.

7. ________________: Phenobarbital, the sedative chloral hydrate, and the anesthetic methohexital are included in this group.

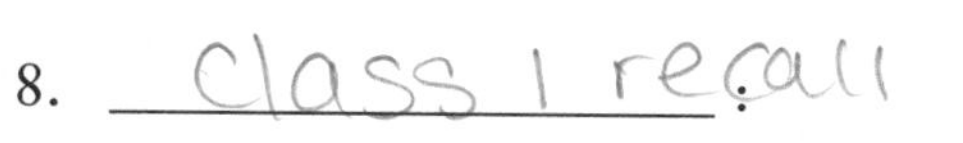

8. ________________: When there is a strong likelihood that the product will cause serious adverse effects or death.

IDENTIFY

Identify the required elements on this manufacturer's bottle label by answering in the space beneath the question.

NDC 60951-602-70 NSN 6505-01-082-5509

Each tablet contains:
Oxycodone Hydrochloride, USP . . . 5 mg*
Acetaminophen, USP . . . . 325 mg
*5 mg oxycodone HCl is equivalent to 4.4815 mg of oxycodone.

Usual Dosage: See package insert.

Store at 20° to 25°C (68° to 77°F).
[see USP Controlled Room Temperature].

This is a bulk package not intended for dispensing.

Dispense in a tight, light-resistant container as defined in the USP, with a child-resistant closure (as required).

DEA ORDER FORM REQUIRED.

Manufactured for:
Qualitest Pharmaceuticals
Huntsville, AL 35811

By: Vintage Pharmaceuticals
Huntsville, AL 35811

ENDOCET® CII
(oxycodone and acetaminophen tablets, USP)

5 mg/325 mg

Endo 602

Multiple strengths: Do not dispense unless strength is stated.

100 TABLETS R_x only

Rev. 9/11 R0
8083396 6807

Lot: ELB089A
Exp: 06/2019

N 3 60951-602-70 9

Endocet® label courtesy Endo Pharmaceuticals Inc.

1. In what kind of container should this medication be dispensed?
   tight + light resistance
2. Who is the manufacturer of this drug product?
   Qualitest Pharm
3. What is the product's brand name?
   endocet
4. What is the product's generic name?
   oxycodone + acetaminophen
5. What is the drug form?
   tablet
6. What are the active ingredients?
   oxycodone + HCL, acetaminophen
7. What control level is the drug product?
   C-2
8. What are the storage requirements?
   20-25°C
9. What is the expiration date?
   06/2019

IN THE WORKPLACE

This MedWatch form 3500 is used to report adverse drug events to the FDA.

U.S. Department of Health and Human Services

MEDWATCH

The FDA Safety Information and Adverse Event Reporting Program

For VOLUNTARY reporting of adverse events, product problems and product use errors

Page 1 of 3

Form Approved: OMB No. 0910-0291, Expires: 9/30/2018
See PRA statement on reverse.

FDA USE ONLY
Triage unit sequence #
FDA Rec. Date

Note: For date prompts of "dd-mmm-yyyy" please use 2-digit day, 3-letter month abbreviation, and 4-digit year; for example, 01-Jul-2015.

A. PATIENT INFORMATION

1. **Patient Identifier** — In Confidence
2. **Age** ______ ☐ Year(s) ☐ Month(s) ☐ Week(s) ☐ Day(s) — or Date of Birth (*e.g., 08 Feb 1925*) __-___-____
3. **Sex** ☐ Female ☐ Male
4. **Weight** ______ ☐ lb ☐ kg

5.a. **Ethnicity** (*Check single best answer*) ☐ Hispanic/Latino ☐ Not Hispanic/Latino

5.b. **Race** (*Check all that apply*) ☐ Asian ☐ American Indian or Alaskan Native ☐ Black or African American ☐ White ☐ Native Hawaiian or Other Pacific Islander

B. ADVERSE EVENT, PRODUCT PROBLEM

1. **Check all that apply**
☐ **Adverse Event** ☐ **Product Problem** (*e.g., defects/malfunctions*)
☐ **Product Use Error** ☐ **Problem with Different Manufacturer of Same Medicine**

2. **Outcome Attributed to Adverse Event** (*Check all that apply*)
☐ Death *Include date (dd-mmm-yyyy):* __-___-____
☐ Life-threatening ☐ Disability or Permanent Damage
☐ Hospitalization – initial or prolonged ☐ Congenital Anomaly/Birth Defects
☐ Other Serious (Important Medical Events)
☐ Required Intervention to Prevent Permanent Impairment/Damage (Devices)

3. **Date of Event** (*dd-mmm-yyyy*) __-___-____
4. **Date of this Report** (*dd-mmm-yyyy*) __-___-____

5. **Describe Event, Problem or Product Use Error**

6. **Relevant Tests/Laboratory Data, Including Dates**

7. **Other Relevant History, Including Preexisting Medical Conditions** (*e.g., allergies, pregnancy, smoking and alcohol use, liver/kidney problems, etc.*)

C. PRODUCT AVAILABILITY

2. **Product Available for Evaluation?** (*Do not send product to FDA*)
☐ Yes ☐ No ☐ Returned to Manufacturer on (*dd-mmm-yyyy*) __-___-____

D. SUSPECT PRODUCTS

1. **Name, Manufacturer/Compounder, Strength** (*from product label*)

#1 – Name and Strength	#1 – NDC # or Unique ID
#1 – Manufacturer/Compounder	#1 – Lot #
#2 – Name and Strength	#2 – NDC # or Unique ID
#2 – Manufacturer/Compounder	#2 – Lot #

3. **Dose or Amount** | **Frequency** | **Route**

	Dose or Amount	Frequency	Route
#1			
#2			

4. **Dates of Use** (*From/To for each*) (*If unknown, give duration, or best estimate*) (*dd-mmm-yyyy*)
#1
#2

5. **Diagnosis or Reason for Use** (*indication*)
#1
#2

6. **Is the Product Compounded?**
#1 ☐ Yes ☐ No
#2 ☐ Yes ☐ No

7. **Is the Product Over-the-Counter?**
#1 ☐ Yes ☐ No
#2 ☐ Yes ☐ No

8. **Expiration Date** (*dd-mmm-yyyy*)
#1 __-___-____ #2 __-___-____

9. **Event Abated After Use Stopped or Dose Reduced?**
#1 ☐ Yes ☐ No ☐ Doesn't apply
#2 ☐ Yes ☐ No ☐ Doesn't apply

10. **Event Reappeared After Reintroduction?**
#1 ☐ Yes ☐ No ☐ Doesn't apply
#2 ☐ Yes ☐ No ☐ Doesn't apply

E. SUSPECT MEDICAL DEVICE

1. **Brand Name**
2. **Common Device Name**
2b. **Procode**
3. **Manufacturer Name, City and State**
4. **Model #** | **Lot #**
Catalog # | **Expiration Date** (*dd-mmm-yyyy*) __-___-____
Serial # | **Unique Identifier (UDI) #**
5. **Operator of Device** ☐ Health Professional ☐ Lay User/Patient ☐ Other
6. **If Implanted, Give Date** (*dd-mmm-yyyy*) __-___-____
7. **If Explanted, Give Date** (*dd-mmm-yyyy*) __-___-____
8. **Is this a single-use device that was reprocessed and reused on a patient?** ☐ Yes ☐ No
9. **If Yes to Item 8, Enter Name and Address of Reprocessor**

F. OTHER (CONCOMITANT) MEDICAL PRODUCTS

Product names and therapy dates (*Exclude treatment of event*)

G. REPORTER (*See confidentiality section on back*)

1. **Name and Address**
Last Name: First Name:
Address:
City: State/Province/Region:
Country: ZIP/Postal Code:
Phone #: Email:

2. **Health Professional?** ☐ Yes ☐ No
3. **Occupation**
4. **Also Reported to:** ☐ Manufacturer/Compounder ☐ User Facility ☐ Distributor/Importer
5. **If you do NOT want your identity disclosed to the manufacturer, please mark this box:** ☐

PLEASE TYPE OR USE BLACK INK

FORM FDA 3500 (10/15) Submission of a report does not constitute an admission that medical personnel or the product caused or contributed to the event.

IN THE WORKPLACE ACTIVITIES

1. Working in pairs, read about the drug recall process at www.fda.gov/ForConsumers/ConsumerUpdates/ucm049070.htm. Discuss with your partner the recall classifications and the FDA-regulated products that can be subject to recalls.

2. Working in pairs, simulate the following situation. A patient who operates dangerous machinery every day brings in a prescription for medication known to cause sedation. The patient tells the pharmacy technician that he needs to take the medication while he's at work. What would you do?

CHOOSE THE BEST ANSWER

Answers are at the end of the book.

1. The ________________ defined what drugs require a prescription.
   a. 1970 Poison Prevention Packaging Act
   b. 1962 Kefauver-Harris Amendment
   c. Sherley Amendment
   d. 1951 Durham-Humphrey Amendment

2. A need for tighter drug regulations from the thalidomide lesson led to the
   a. Kefauver-Harris Amendment.
   b. Durham-Humphrey Amendment.
   c. Food and Drug Act of 1906.
   d. Food Drug and Cosmetic Act.

3. Legend drugs should have the legend ________________ on the manufacturer's label.
   a. "Federal law prohibits transfer of this prescription"
   b. "Store at room temperature"
   c. "For external use only"
   d. "RX only"

4. In clinical trials, the testing is done
   a. on mice.
   b. on people.
   c. in vitro.
   d. on dogs.

5. The main purpose of phase II clinical trials is
   a. efficacy.
   b. dosage.
   c. safety.
   d. economics.

6. The monthly sales limit of pseudoephedrine base is
   a. 7.5 g per household.
   b. 7.5 g per purchaser.
   c. 7.5 g per transaction.
   d. 7.5 g per drug per household.

7. The national drug code (NDC) is assigned by the
   a. FDA.
   b. DEA.
   c. CDER.
   d. manufacturer.

8. DEA form __________ is used to order Schedule II controlled substances.
   a. 41
   b. 106
   c. 222
   d. 224

9. Legal agreements that have duties associated with them.
   a. Negligence
   b. Injunctions
   c. Torts
   d. Contracts

10. A branch of philosophy that helps determine what should be done in a principled sense.
    a. Ethics
    b. Autonomy
    c. Beneficence
    d. Justice

STUDY NOTES

Use this area to write important points you'd like to remember.

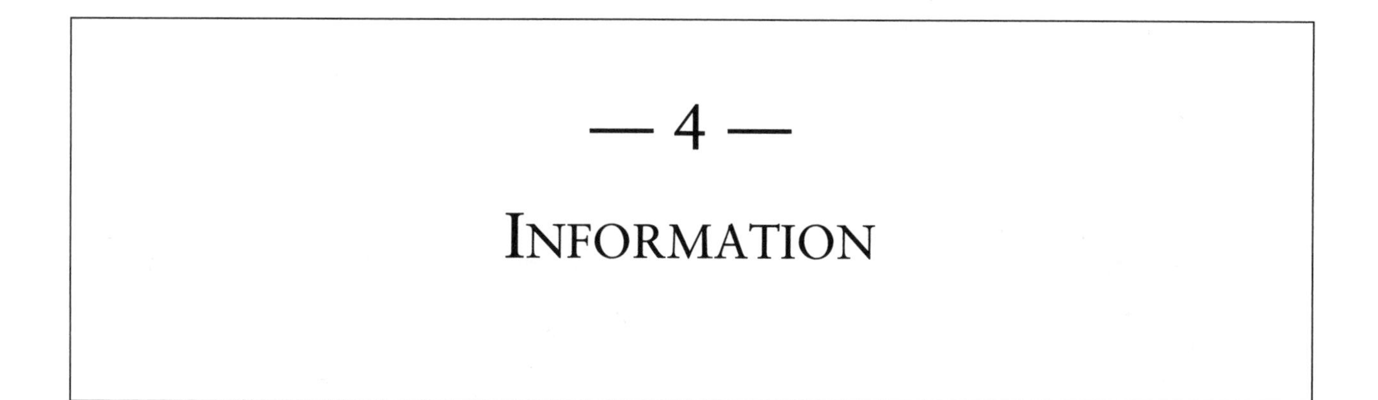

KEY CONCEPTS

Test your knowledge by covering the information in the right-hand column.

primary literature	Original reports of clinical trials, research, and case studies; used for the most updated information.
secondary literature	Indexing or abstracting services.
tertiary literature	Summary information based on primary literature.
abstracting services	Services that summarize information from various primary sources for quick reference.
Safety Data Sheets (SDSs)	OSHA required information for handling hazardous chemicals.
state regulations	Many states have Pharmacy Statutes or State Board of Pharmacy Rules and Regulations that require pharmacies to maintain specific professional literature references.
AHFS Drug Information	Comprehensive, evidence-based reference. It groups drug monographs by therapeutic use.
Physicians' Desk Reference	An annual publication that contains information similar to pharmaceutical manufacturers' drug package inserts.
***Drug Facts and Comparisons* (DFC)**	A preferred reference for comprehensive and timely drug information, containing information about prescription and OTC products.
Handbook on Injectable Drugs	A collection of monographs on commercially available parenteral drugs that include preparation, storage administration, compatibility, and stability of injectable drugs.
RED BOOK	Guide to products and prices, providing annual price lists of drug products including manufacturer, package size, strength, and wholesale and retail prices.

"Orange Book" — The common name for the FDA's *Approved Drug Products with Therapeutic Equivalence Evaluations*.

Merck Index — Information on chemicals, drugs, and biologicals including names, chemical structures, and physical and toxicity data.

Martindale: The Complete Drug Reference — Formerly known as "Martindale: The Extra Pharmacopoeia," contains information on drugs in clinical use internationally.

King Guide to Parenteral Admixtures — This reference provides information on injectable drug compatibility and stability of injectable drugs.

Internet — A "supernetwork" of many networks from around the world all connected to each other by telephone lines, and all using a common "language."

mobile medical apps — Applications for mobile devices including smartphones and iPads that can be used for accessing drug information.

STUDY NOTES

Use this area to write important points you'd like to remember.

FILL IN THE KEY TERM

Answers are at the end of the book.

Drug Facts and Comparisons	Lexicomp	*RED BOOK*
Handbook on Injectable Drugs	*Martindale: The Complete Drug Reference*	secondary literature
Handbook of Nonprescription Drugs	Safety Data Sheets (SDSs)	tertiary literature
King Guide to Parenteral Admixtures	"Orange Book"	*United States Pharmacopeia–National Formulary (USP–NF)*
	Physicians' Desk Reference	
	primary literature	

1. 1st literature: Original reports of clinical and other types of research projects and studies.
2. 3rd literature: Condensed works based on primary literature, such as textbooks, monographs, etc.
3. Red book: Reference with updated AWPs and NDCs.
4. SDS: OSHA required information for handling hazardous chemicals.
5. 2 literature: General reference works based upon primary literature sources.
6. lexicomp: Subscription-based online databases and tools.
7. Drug facts and compansons: A preferred reference for comprehensive and timely drug information, containing information about prescription and OTC products.
8. physican Desk refe: Provides information similar to drug package inserts.
9. Handbook of injectable: A collection of monographs on commercially available parenteral drugs that include topics such as preparation, storage, administration, compatibility, and stability.
10. King guide: Information on injectable drug compatibility and stability.
11. Martindale: Formerly called "The Extra Pharmacopeia," contains international drug monographs.
12. orange book: Used to determine therapeutic equivalence of brand and generic drugs.
13. Handbook of non pres: Quick access to OTC information published by APhA.
14. USP-NF: Compendia of drug information. Chapters numbered below <1,000> are enforceable by the Food and Drug Administration.

TRUE/FALSE

Indicate whether the statement is true or false in the blank. Answers are at the end of the book.

__T__ 1. Primary literature provides the largest amount of and most current source of information.

__T__ 2. Abstracting services are considered secondary literature.

__T__ 3. *Drug Facts and Comparisons* provides information on prescription and OTC drug products, using comparative tables of therapeutic groups.

__F__ 4. *AHFS* is a collection of monographs on parenteral drugs.

__F__ 5. Micromedex is only available in the print edition.

__T__ 6. The "Orange Book" is available online.

__T__ 7. The *RED BOOK* provides pricing information.

__T__ 8. PubMed indexes primary medical literature.

__F__ 9. Medline is an example of primary literature.

__T__ 10. Clinical Pharmacology is accepted by all 50 state boards of pharmacy as an official drug reference.

__T__ 11. The National Association of Boards of Pharmacy (NABP) developed an accreditation program for pharmacy practices on the Internet.

EXPLAIN WHY

Explain why these statements are true or important. Check your answers in the text. Discuss any questions you may have with your Instructor.

1. Why is primary literature important?

2. Why is continuing education important?

3. Why should technicians be familiar with pharmaceutical information sources?

4. Why would a health-care professional choose an MMA that is not free?

IN THE WORKPLACE ACTIVITY

1. Use the SDS Online search engine www.uspmsds.com/msds/controller to find safety data sheets for the following products: normal saline, sterile water for injection, allopurinol, hydrochlorothiazide, methotrexate.

CHOOSE THE BEST ANSWER

Answers are at the end of the book.

1. _____________________ literature contains condensed works based on primary literature, such as textbooks, monographs, etc.
   a. Orange
   b. Secondary
   c. Tertiary
   d. Abstract

2. OSHA requires pharmacies to have _____________ for each hazardous chemical on hand.
   a. manufacturer sheets for documentation of safety
   b. Safety Data Sheets (SDSs)
   c. mixture safety documentation sheets
   d. manufacturer's safety documentation sheets

3. _________________ summarize(s) information from primary sources for quick reference.
   a. The *PDR*
   b. *American Drug Index*
   c. *US Pharmacist*
   d. Abstracting services

4. _____________ is an example of a trade journal.
   a. *America's Pharmacist*
   b. Epocrates
   c. *Pharmacy Law Digest*
   d. *Pharmacy Times*

5. _________________ is a collection of monographs on commercially available parenteral drugs.
   a. The *Merck Index*
   b. The "Orange Book"
   c. *Handbook on Injectable Drugs*
   d. VAERS

6. _____________ is an example of a professional practice journal.
   a. *Today's Technician*
   b. *Pharmacy Times*
   c. *US Pharmacist*
   d. *The Medical Letter*

7. *The Pharmacist's Letter* is an example of a(n)
   a. MMA.
   b. trade journal.
   c. professional practice journal.
   d. newsletter.

8. The most useful reference for dosing medicines for children
   a. *Drugs in Pregnancy and Lactation*
   b. Lexicomp
   c. Micromedex
   d. *Pediatric and Neonatal Dosage Handbook*

9. The most useful reference for dosing medications for the geriatric population.
   a. Micromedex
   b. Lexicomp
   c. *Geriatric Dosage Handbook*
   d. Epocrates

10. The best reference to research foreign drugs.
    a. *Martindale*
    b. Micromedex
    c. Epocrates
    d. Lexicomp

11. Organization that provides a Medication Error Reporting Program (MERP).
    a. MEDMARX
    b. ISMP
    c. USP–NF
    d. OSHA

12. The *Occupational Outlook Handbook* is available from
    a. the Bureau of Labor Statistics (BLS).
    b. the Pharmacy Technician Educators Council (PTEC).
    c. the American Society of Health System Pharmacists (ASHP).
    d. the American Pharmacists Association (APhA).

13. The organization that helps guide users to reliable web content.
    a. ASHP
    b. APhA
    c. Health on the Net Foundation
    d. Verified Internet Pharmacy Practice Sites

14. The National Institute of Health (NIH) free website for patients.
    a. Medscape
    b. Epocrates
    c. Lexicomp
    d. MedlinePlus

STUDY NOTES

Use this area to write important points or Web addresses you'd like to remember.

— 5 —

TERMINOLOGY

KEY CONCEPTS

Test your knowledge by covering the information in the right-hand column.

terminology	Much of medical science is made up of a small number of root words, suffixes, and prefixes that originated from either Greek or Latin words.
root word	The base component of a term that gives it a meaning that may be modified by other components.
prefix	A modifying component of a term located before the other components of the term.
suffix	A modifying component of a term located after the other components of the term.
combining vowel	Combining vowels are used to connect the prefix, root, or suffix parts of the term.
integumentary system	The body's first line of defense, acting as a barrier against disease and other hazards.
skeletal system	Protects soft organs and provides structure and support for the body's organ systems.
muscular system	The body contains more than 600 muscles, which give shape and movement to it.
nervous system	The body's system of communication. The neuron (nerve cell) is its basic functional unit.
cardiovascular system	Distributes blood throughout the body using blood vessels called arteries, capillaries, and veins.
lymph and blood systems	The center of the body's immune system. Lymphocytes are white blood cells that helps the body defend itself against bacteria and diseased cells.

respiratory system	Brings oxygen into the body through inhalation and expels carbon dioxide gas through exhalation.
gastrointestinal (GI) tract	Contains the organs that are involved in the digestion of foods and the absorption of nutrients.
urinary tract	The primary organ is the kidney, which filters the blood for unwanted material and makes urine.
endocrine system	Consists of the glands that secrete hormones (chemicals that assist in regulating body functions).
female reproductive system	Produces hormones (estrogen, progesterone), controls menstruation, and provides for childbearing.
male reproductive system	Produces sperm and secretes the hormone testosterone.
senses: hearing	The sense of hearing, as well as the maintenance of body equilibrium, is performed by the ear.
senses: sight	The eyelids protect the eye and assist in its lubrication. The conjunctiva is the blood-rich membrane between the eye and the eyelid.
drug classifications	A grouping of a number of drugs that have some properties in common. The same steps in interpreting other medical science terminology can be used to interpret drug classification names.
medical abbreviation	Many medical terms are abbreviated for communication and record keeping purposes.

Study Notes

Use this area to write important points you'd like to remember.

ORGAN SYSTEM ROOTS

INTEGUMENTARY SYSTEM

adip	fat
cutane	skin
derm(at)	skin
hist	tissue
ichthy	dry; scaly
kerat	hard
mast	breast
melan	black
necr	death (of cells, etc.)
onych	nail
prurit, psor	itching

SKELETAL SYSTEM

arthr	joint
carp	wrist
crani	skull
dactyl	finger or toe
fibul	small, lower leg bone
humer	upper arm bone
lord	curve
oste	bone
patell	kneecap
ped, pod	foot
phalang	bones of fingers and toes
rachi	vertebrae
scoli	crooked, bent
spondyl	vertebrae
stern	sternum, breastbone
tars	ankle

MUSCULAR SYSTEM

burs	bursa
chondr	cartilage
fasci	fiber separating muscles
fibr	fiber
leiomy	smooth muscle
my	muscle
rhabdo	rod
tendin	tendon

NERVOUS SYSTEM

alges	pain
cerebr	cerebrum
encephal	brain
esthes	sensation
lex	word, phrase
mening	meninges
ment	mind
myel	spinal cord
neur	nerve
phas	speech
psycho	mind
somat	body

CARDIOVASCULAR SYSTEM

aneur	widening
angi	vessel
aort	aorta
arter(i)	artery
card(i)	heart
coron	heart
cyt(o)	cell
embol	embolus
oxy	oxygen
pector	chest
phleb	vein
sten(o)	narrowed
thromb	clot
vas(cu)	blood vessel
ven	vein

LYMPH AND BLOOD SYSTEMS

aden	gland
bacter	bacteria
cyt	cell
hemo, hemat	blood
leuk	white
lymph	lymph
phleb	vein
sepsis	to putrify
splen	spleen
thromb	clot
thym	thymus

RESPIRATORY SYSTEM

aer	air
aero	gas
bronch(i)	bronchus
capn	carbon dioxide
cyan	blue
laryng	larynx
nas	nose
ox	oxygen
pector	chest
pneum(on)	lung, air
pulmon	lung
respir	breath
rhin	nose
sinus	sinus
spir	breathe
trache	trachea

GASTROINTESTINAL SYSTEM

chol	bile
col(on)	colon
duoden	duodenum
enter	intestine
esophag	esophagus
gastr	stomach
hemat	blood
hepat	liver
herni	hernia
lapar	abdomen
pancreat	pancreas
pepsia	digestion
phag	swallow

URINARY TRACT

albumin	protein
cyst	bladder
glycos	glucose
keto	ketones
lith	stone
nephr	kidney
ren	kidney
ur	urine; urinary
uresis	urination
ureter	ureter
urethr	urethra
vesic	bladder

ENDOCRINE SYSTEM

aden	gland
adren(al)	adrenal
crine	to secrete
glyc	sugar
gluc(os)	sugar
lipid	fat
myx	mucos
nephr	kidney
pancreat	pancreas
plas	development
somat	body
tetan	tetanus
thym	thymus
thyr, thyroid	thyroid

FEMALE REPRODUCTIVE SYSTEM

cervic	cervix
condyle	knob, knuckle
eclamps	shining forth
gynec	woman
hyster	uterus
lact	milk
mamm	breast
mast	breast
men	menstruation
metr(i)	uterus
ovari	ovary
salping	fallopian tube
toc	birth
uter	uterine
vagin	vagina

MALE REPRODUCTIVE SYSTEM

andr	male
balan	glans penis
crypt	hidden
orch, orchid	testis
prostat	prostate gland
semin	semen
sperm	sperm
test	testicle
varic	varicose veins
vas	vessel, duct
vesicule	seminal vesicle

HEARING

acous	hearing
acusis	hearing condition
audi	hearing
cerumin	waxlike
labyrinth	inner ear
myring	eardrum
ot	ear
salping	eustachian tube
tympan	eardrum

SIGHT

ambly	dim, dull
blephar	eyelid
conjuctiv	conjunctiva
corne	cornea
glauc	gray
irid, ir	iris
lacrim	tear duct
ocul	eye
morph	shape, structure
ophthalm	eye
opia	vision
opt	eye
retin	retina
stigmat	point(ed)

COMMON PREFIXES

a	without
ambi	both
an	without
ante	before
anti	against
bi	two or both
brady	slow
chlor	green
circum	around
cirrh	yellow
con	with
contra	against
cyan	blue
dia	across or through
dipl	double
dis	separate from or apart
dys	painful, difficult
ec	away or out
ecto	outside
en	in, within
endo	within
epi	upon
erythr	red
eso	inward
eu	good or normal
exo	outside
hemi	half
heter	different
hyper	above or excessive
hypo	below or deficient
im	not
immun	safe, protected
in	not
infra	below or under
inter	between
intra	within
iso	equal
leuk	white
lip	fat
macro	large
medi	middle
melan	black
meso	middle
meta	beyond, after, changing
micro	small
mid	middle
mono	one
multi	many
neo	new
pachy	heavy, thick
pan	all
para	alongside or abnormal
peri	around
polio	gray
poly	many
post	after
pre	before
pro	before
pseudo	false
purpur	purple
quadri	four
re	again or back
retro	after
rube	red
semi	half
sub	below or under
super	above or excessive
supra	above or excessive
sym, syn	with
tachy	fast
trans	across, through
tri	three
ultra	beyond or excessive
uni	one
xanth	yellow
xero	dry

COMMON SUFFIXES

Suffix	Meaning
ac, al, ar, ary	pertaining to
algia	pain
ar	pertaining to
asthenia	without strength
blast	immature cell
cele	pouching or hernia
cyesis	pregnancy
cynia	pain
eal	pertaining to
ectasis	expansion or dilation
ectomy	removal
edema	swelling
ema	condition
emesis	vomiting
emia	blood condition
esthenia	lack of sensation
genic	origin or production
gram	record
graphy	recording process
ia	condition of
iasis	condition, formation of
iatry	treatment
ic	pertaining to
icle	small
ism	condition of
itis	inflammation
ium	tissue or structure
lith	stone, calculus
logy	study of
lysis	breaking down, dissolution
malacia	softening
megaly	enlargement
metry	measuring process
myc	fungus
oid	resembling
ole	small
ology	study of
oma	tumor
opia, opsia	vision
orexia	appetite
ose	full of, pertaining to
osis	abnormal condition
osmia	smell
ous	pertaining to
paresis	partial paralysis
pathy	disease
penia	decrease
pepsia	digestion
phagia	swallowing
phasia	speech
philia	attraction for
phobia	fear
plasia	formation, development
plasty	surgical repair, reconstruction
plegia	paralysis, stroke
pnea	breathing
polesis	formation, production
rrhage	to burst forth
rrhea	discharge
sclerosis	hardness
scopy	examination
spasm	involuntary contraction
stasis	stop or stand
stomy	new opening
tic	pertaining to
tocia	childbirth, labor
tomy	incision
toxic	poison
tropea	to turn
tropic	stimulate
trophy	nutrition, growth
ula, ule	small
y	condition, process of

COMMON MEDICAL ABBREVIATIONS

ADD	attention deficit disorder	HIV	human immunodeficiency virus
AFib	atrial fibrillation	HR	heart rate
AIDS	acquired immunodeficiency syndrome	HRT	hormone replacement therapy
AMS	altered mental state	HTN	hypertension
ASAP	as soon as possible	IO, I/O	fluid intake and output
AV	atrial-ventricular	IOP	intraocular pressure
BM	bowel movement	KVO	keep veins open
BMI	body mass index	LBW	low birth weight
BP	blood pressure	LOC	loss of consciousness
BS	blood sugar	MI	myocardial infarction
BSA	body surface area	MDI	metered dose inhaler
CA	cancer	NKA	no known allergies
CABG	coronary artery bypass graft	NPO	nothing by mouth
CAD	coronary artery disease	NVD	nausea, vomiting, diarrhea
CBC	complete blood count	OM	otitis media
CHD	coronary heart disease	OR	operating room
CHF	congestive heart failure	PMH	past medical history
COPD	chronic obstructive pulmonary disease	PUD	peptic ulcer disease
CVA	cerebrovascular accident (stroke)	PVD	peripheral vascular disease
DAW	dispense as written	RA	rheumatoid arthritis
D/C	discontinue	RBC	red blood count or red blood cell
DM	diabetes mellitus	SBP	systolic blood pressure
DOB	date of birth	SOB	shortness of breath
DUR	drug utilization review	STD	sexually transmitted diseases
Dx	diagnosis	T	temperature
EC	enteric coated	T&C	type and cross-match (blood)
ECG, EKG	electrocardiogram	TB	tuberculosis
ENT	ears, nose, throat	TEDS	thrombo-embolic disease stockings
ER	emergency room	TPN	total parenteral nutrition
ESRD	end stage renal disease	Tx	treatment
FH	family history	U	units
GERD	gastroesophageal reflux disease	U/A	urinalysis
GI	gastrointestinal	URD	upper respiratory diseases
Gtt	drop	UTI	urinary tract infection
HA	headache	VS	vital signs
HBP	high blood pressure	WBC	white blood count or white blood cell
HD	hemodialysis	WT	weight
HDL	high density lipoprotein		

FILL IN THE KEY TERM

Use these key terms to fill in the correct blank. Answers are at the end of the book.

aphagia	**dysuria**	**hernia**	**pachyderm**
arthritis	**eczema**	**hyperglycemia**	**phlebitis**
blepharitis	**encephalitis**	**hypertrophy**	**phlebotomy**
bronchitis	**endocrine**	**hypotonia**	**prostatolith**
cardiomyopathy	**endometriosis**	**lordosis**	**tendinitis**
colitis	**fibromyalgia**	**lymphoma**	**thrombosis**
cystitis	**hematoma**	**neuralgia**	**vasectomy**
dyspepsia	**hemophilia**	**parathyroid**	

1. ____________________ : Decrease in muscle tone.
2. ____________________ : Chronic pain in the muscles.
3. ____________________ : Inflammation of the brain.
4. ____________________ : Pertaining to glands that secrete hormones into the bloodstream.
5. ____________________ : Disease of the heart muscle.
6. ____________________ : Next to the thyroid gland.
7. ____________________ : Inflammation of the joint.
8. ____________________ : High blood sugar.
9. ____________________ : Inability to swallow.
10. ____________________ : Condition of indigestion.
11. ____________________ : Inflamed or irritable colon.
12. ____________________ : Process of increase in muscle size.
13. ____________________ : Protrusion of organ or tissue.
14. ____________________ : Painful urination.
15. ____________________ : Puncture of vein to draw blood.
16. ____________________ : Blood clots in the vascular system.
17. ____________________ : A collection of blood, often clotted.
18. ____________________ : A disease in which the blood does not clot normally.
19. ____________________ : Lymphatic system tumor.
20. ____________________ : Chronic skin inflammation.
21. ____________________ : Inflammation of a tendon.
22. ____________________ : Severe pain in a nerve.
23. ____________________ : Abnormal thickness of skin.
24. ____________________ : Abnormal growth of uteral tissue within the pelvis.
25. ____________________ : Removal of a section of the vas deferens.
26. ____________________ : A prostate stone.
27. ____________________ : Inflammation of bronchial membranes.
28. ____________________ : Inflammation of a vein.
29. ____________________ : Inflammation of the bladder.
30. ____________________ : Forward curve of spine.
31. ____________________ : Inflammation of the eyelids.

IN THE WORKPLACE

Read through the story of a woman named Mary Ellen and fill in the blanks.

Mary Ellen presents to the ER today complaining of nausea, vomiting. She has developed (1)_______________ due to her loss of appetite and finds it hard to swallow, a term known as (2)_______________. The doctor brings her in for an exam and finds that her colon is inflamed, a condition known as (3) _______________. He explains that this is the reason why she has been experiencing these gastrointestinal issues.

Mary Ellen states that she has a history of GERD (gastroesophageal reflux disease) and (4)_______________ (a pancreatic disorder of insufficient insulin production) and notes that her esophagus feels swollen (also known as (5) _______________). She has been so stressed out by her loss of appetite that she has begun to lose hair! This condition of hair loss, (6) _______________, has been really worrying her. You can tell that she is nervous by the excessive sweating, also known as (7) _______________. You try to calm her down by telling her that there is a solution to her problem. You decide to measure her blood glucose while you are at it and discover that she has high blood sugar, a condition known as (8) _______________.

You ask her about her past medical history and medication history. She tells you that she scraped her knee on the sidewalk a few weeks ago, but never found the time to come in to get it checked out. So you examine her knee and see a large, red, inflamed, infected wound. You worry about (9)_______________, the presence of bacteria in the bloodstream. You quickly talk to the doctor and he orders a Complete Blood Count (CBC) with differential. He specifically wants to see the white blood cells (also called (10) _______________) and lymphocytes. He is not sure if Mary Ellen has (11) _______________, a systemic blood infection. After the CBC comes back, it turns out that she does not have a systemic infection and that it is only (12) _______________, infection of the skin. She seems relieved already.

At her follow-up, a few weeks later, Mary Ellen tells you that she has gotten her appetite back! She has even joined a soccer team to keep her busy and healthy. She tells you that she plays every night for 3 hours. She says she is super proud of her new muscular legs, but that she feels (13)_______________, muscle pain. You take a look at her legs to make sure everything is all right and notice that her muscles have increased in size, a term known as (14) _______________. She tells you that it is from all the practice that she has been getting each night. She also tells you that she does not want them to undergo (15) _______________, the process of shrinking muscle size. Worried, you recommend her to take breaks more often and lessen the intensity of her exercise each day. Otherwise, over time she may begin to feel (16) _______________, the breaking down of motor control.

She also tells you that she has been experiencing (17) _______________, a temporary failure to breathe. She notes that this happens at night after soccer practice. During these "episodes" she feels (18) _______________, chest pain. You refer her to a pulmonary specialist immediately. In the meantime, you recommend that she stops playing soccer for a while.

You ask Mary Ellen if there is anything else that you can do for her at the visit. She bashfully says, "Well yes, I have recently missed my period for 4 months now. I took 3 pregnancy tests, but they have all come out negative." The condition that she describes is known as (19) _______________, an absence of menstruation. She asks if this is something that she should worry about. You refer her to an OB-GYN for further evaluation.

Additionally, she notes that she has been feeling discomfort/pain, a term known as (20)_______________, when she urinates. She feels a stinging/burning sensation. You order a urinalysis for her. You find that she has (21) _______________ (protein in the urine), (22)_______________ (glucose in the urine), (23) _______________ (blood in the urine), and (24)_______________ (ketone bodies in the urine). This suggests that she has a UTI (urinary tract infection). The doctor writes her a prescription for nitrofurantoin for 7 days. You counsel her on the side effects and importance of taking it properly every day for the full course. Mary Ellen is happy that she will feel some relief soon.

You ask her if she has had problems with her heart in the past. She responds that she has an abnormally slow heart rate, also known as (25)____________________. Her father and mother both have (26) _______________ (high blood pressure). They also have (27) _______________ (hardening of the artery) and cholesterol problems. You measure her heart rate and it comes out on the lower end of normal. She looks relieved to have received this news.

IN THE WORKPLACE ACTIVITIES

1. Working in pairs, practice pronouncing the following words in front of your partner. The partner should provide a critique of the pronunciation of the words. Antacid, antianginal, anticoagulant, anticonvulsant, antidepressant, antidiarrheal, anti-hyperlipidemic. Use the National Library of Medicine online resource MedlinePlus (https://www.nlm.nih.gov/medlineplus/) as a guide for pronunciation.
2. Working in pairs and using the information on pages 42–45 on roots, prefixes and suffixes, determine the meaning of the following terms: Aneurysm, Angioplasty, Cardiograph, Phlebitis, Gastritis, Anorexia, Neuralgia, Parathyroid, Tendinitis, Fibromyalgia, Dyspepsia.

CHOOSE THE BEST ANSWER

Answers are at the end of the book.

1. The system of medical and pharmaceutical nomenclature is made of these four elements:
   a. prefixes, suffixes, root words, and combining vowels.
   b. prefixes, suffixes, key words, and combining vowels.
   c. prefixes, suffixes, key words, and combining consonants.
   d. prefixes, suffixes, root words, and combining consonants.

2. The root word "partum" means
   a. skin.
   b. bring forth.
   c. chest.
   d. lung.

3. Hematemesis is the vomiting of
   a. partially digested food.
   b. clear colorless liquid.
   c. blood.
   d. kidney.

4. The suffix "paresis" means
   a. treatment.
   b. tissue or structure.
   c. formation.
   d. partial paralysis.

5. Sublingual means
   a. narrowing of the tongue.
   b. inflammation of the tongue.
   c. under the tongue.
   d. without strength.

6. Prostatitis means
   a. a prostate stone.
   b. inability to produce semen.
   c. inflammation of the testes.
   d. inflammation of the prostate.

7. The major functional organ of the urinary tract is the
   a. liver.
   b. kidney.
   c. bladder.
   d. intestine.

8. A keratosis is
   a. a skin inflammation.
   b. a fungal infection of the nails.
   c. an area of increased hardness.
   d. softening of the skin.

9. The suffix "emia" means
   a. tissue.
   b. pain.
   c. blood condition.
   d. recording process.

10. The suffix "stomy" means
   a. condition of.
   b. new opening.
   c. fear of.
   d. pain.

11. Arteriosclerosis means
   a. disease of the heart.
   b. concerning heart muscle.
   c. hardening of the arteries.
   d. clotting of blood.

12. The suffix "rrhage" means
   a. to burst forth.
   b. to discharge.
   c. around.
   d. origin or production.

13. The prefix "pseudo" means
   a. halt.
   b. false.
   c. around.
   d. with.

STUDY NOTES

Use this area to write important points you'd like to remember.

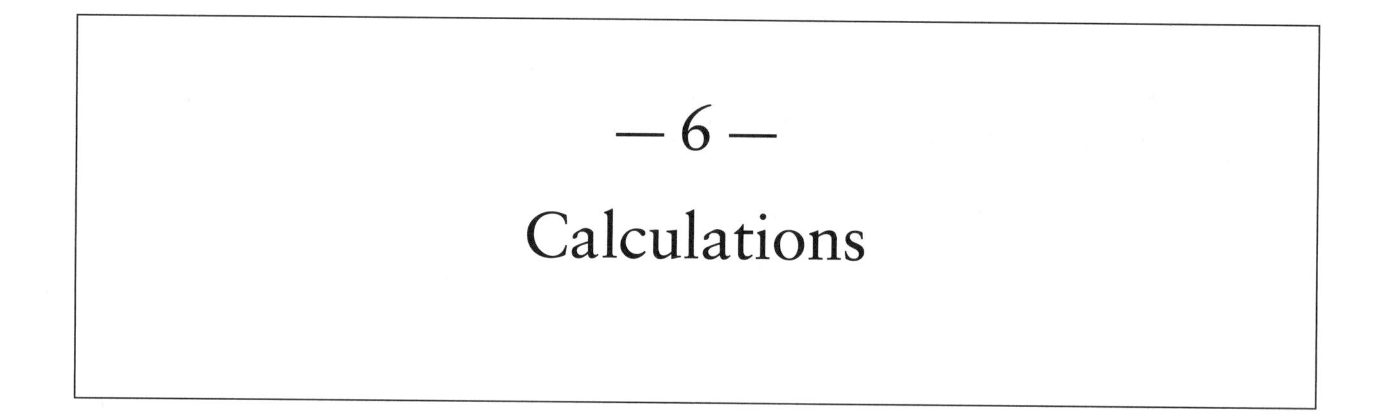

— 6 —

Calculations

ROMAN NUMERALS

Roman numerals can be capital or lowercase letters, and are:

ss = 1/2
I or i = 1
V or v = 5
X or x = 10
L or l = 50
C or c = 100
D or d = 500
M or m = 1,000

RULES FOR POSITIONAL NOTATION

1. When Roman numerals are repeated, their values are added together.
   Example: XXX = 10 + 10 + 10 = 30
2. Roman numerals cannot be repeated more than three times. **Example:** III, XXX, etc.
3. V cannot be repeated. **Example:** V is a numeral but VV is not a numeral.
4. When a smaller numeral follows a larger numeral, their values are added together.
   Example: VI = 5 + 1 = 6
5. When a smaller numeral is followed by a larger numeral, the smaller numeral is subtracted from the larger numeral. **Example:** IV = 5 – 1 = 4
   - ➡ Ones (I) may be subtracted only from fives (V) or tens (X).
     Example: IV = 4 and IX = 9
   - ➡ Tens (X) may only be subtracted from fifties (L) and hundreds (C).
     Example: XL = 40 and XC = 90
   - ➡ Hundreds (C) may be subtracted only from five hundreds (D) and thousands (M)
     Example: CD = 400 and CM = 900
6. When a smaller numeral is between two larger numerals, rule 4 is applied first (subtraction), then rule 5 is applied (addition). **Example:** XXIX = 29

Roman Numeral Exercises

Write the following in Roman numerals:

1. 67 LXVII
2. 29 XXXI
3. 41 XLI
4. 108 CVII
5. 6 VI
6. 98 XCVII
7. 9 IX

Write the following in arabic numbers:

8. XIX 19
9. CIII 103
10. CMM 1900
11. iss 1 1/2
12. XX 20
13. LIV 54

Study Notes

Use this area to write important points you'd like to remember.

CONVERSIONS

Liquid Metric

1 L	=	10 dL	=	1,000 mL
1 dL	=	0.1 L	=	100 mL
1 mL	=	0.001 L	=	0.01 dL

Solid Metric

1 kg	=	1,000 g		
1 g	=	0.001 kg	=	1,000 mg
1 mg	=	0.001 g	=	1,000 mcg
1 mcg	=	0.001 mg		

Avoirdupois

1 lb	=	16 oz
1 oz	=	437.5 gr
1 gr	=	64.8 mg (.0648 g)

Apothecary

1 gal	=	4 qt
1 qt	=	2 pt
1 pt	=	16 fl oz
1 fl oz	=	8 fl dr
1 fl dr	=	60 m

Household

1 tsp	=	5 mL		
1 tbs	=	3 tsp	=	15 mL
1 cup	=	8 fl oz		

Temperature

F temperature = (9/5 times number of degrees C) + 32

C temperature = 5/9 x (number of degrees F - 32)

Conversions Between Systems

1 L	=	33.8 fl oz	1 lb	=	453.59 g
1 pt	=	473.167 mL	1 oz	=	28.35 g
1 fl oz	=	29.57mL	1 g	=	15.43 gr
1 kg	=	2.2 lb	1 gr	=	64.8 mg

CONVERSION EXERCISES

Convert these numbers to decimals:

1. 5/8 0.625
2. 0.2% 0.002
3. 1/5 0.2
4. 2/3 0.67
5. 12% 0.12
6. 38.5% 0.385
7. 1.5% 0.015

Convert these numbers to percents:

8. 0.7 70%
9. 75/100 75%
10. 1.5 150%
11. 1/4 25%
12. 0.04 40%
13. 4/5 80%
14. 0.025 2.5%

RATIO AND PROPORTION

Most of the calculations pharmacy technicians will face on the job or in the certification exam can be performed using the ratio and proportion method.

A ratio states a relationship between two quantities. ➡ $\frac{a}{b}$

A proportion contains two equal ratios. ➡ $\frac{a}{b} = \frac{c}{d}$

When three of the four quantities in a proportion are known, the value of the fourth (x) can be easily solved. ➡ $\frac{x}{b} = \frac{c}{d}$

CONDITIONS FOR USING RATIO AND PROPORTION

1. Three of the four values must be known.
2. Numerators must have the same units.
3. Denominators must have the same units.

STEPS FOR SOLVING PROPORTION PROBLEMS

1. Define the variable and correct ratios.
2. Set up the proportion equation.
3. Establish the x equation.
4. Solve for x.
5. Express the solution in correct units.

Example

If there are 125 mg of a substance in a 500 mL solution, and 50 mg is desired, the amount of solution needed can be determined with this equation:

$$\frac{x \text{ mL}}{50 \text{ mg}} = \frac{500 \text{ mL}}{125 \text{ mg}}$$

multiplying both sides by 50 mg gives:

$$x \text{ mL} = 2{,}500 \text{ mL}/125$$

solving for x gives:

$$x = 200$$

answer: 200 mL of solution are needed.

FLOW RATE

In some settings, the flow rate, or rate of administration for an IV solution, needs to be calculated. This is done using a ratio and proportion equation. Rates are generally calculated in mL/hour, but for pumps used to dispense IV fluids to a patient, the calculation may need to be done in mL/min or gtt/min.

For example, if you have an order for KCl 10 mEq and K Acetate 15 mEq in D5W 1,000 mL to run at 80 mL/hour, you would determine the administration rate in mL/minute as follows:

$$x \text{ mL/1 min} = 80 \text{ mL/60 min}$$

$$x = 80/60 = 1.33$$

To get **drops per minute (gtt/min)**, you must have a conversion rate of drops per mL. For example, if the administration set for the above order delivered 30 drops per mL, you would find the drops per minute as follows:

$$\frac{80 \text{ mL}}{60 \text{ min}} \times \frac{30 \text{ gtt}}{1 \text{ mL}} = \frac{2400 \text{ gtt}}{60 \text{ min}} = 40 \text{ gtt/min}$$

PERCENTS & SOLUTIONS

Percents are used to indicate the amount, or **concentration,** of something in a solution. Concentrations are indicated in terms of weight to volume or volume to volume. The standard units are:

Weight to Volume: grams per 100 milliliters ➡ g/100 mL

Volume to Volume: milliliters per 100 milliliters ➡ mL/100 mL

A PERCENT SOLUTION FORMULA

Technicians find that they often have to convert a solution at one concentration to a solution having a different concentration, especially during the preparation of hyperalimentation, or TPNs. It is possible to make such conversions using a simple proportion equation with these elements:

$$\frac{x \text{ volume wanted}}{\text{want \%}} = \frac{\text{volume prescribed}}{\text{have \%}}$$

MILLIEQUIVALENTS—MEQ

Electrolytes are substances that conduct an electrical current and are found in the body's blood, tissue fluids, and cells. Salts are electrolytes and saline solutions are a commonly used electrolyte solution. The concentration of electrolytes in a volume of solution is measured in units called milliequivalents (mEq). They are expressed as milliequivalents per milliliter or equivalents per liter.

Milliequivalents are a unit of measurement specific to each electrolyte. For example, a 0.9% solution of one electrolyte will have a different mEq value than a 0.9% solution of another because mEq values are based on each electrolyte's atomic weight and electron properties, each of which is different.

If the mEq value of a solution is known, it is relatively easy to mix it with other solutions to get a different mEq volume ratio by using proportions.

EXAMPLE

A solution calls for 5 mEq of an electrolyte that you have in a 1.04 mEq/mL solution. How many mL of it do you need?

x mL/5 mEq = 1 mL/1.04 mEq

$$x \text{ ml} = 5 \,\cancel{\text{mEq}} \text{ times } \frac{1 \text{ mL}}{1.04 \,\cancel{\text{mEq}}} = \frac{5 \text{ mL}}{1.04} = 4.8 \text{ mL}$$

Answer: 4.8 mL of the solution is needed.

ALLIGATION

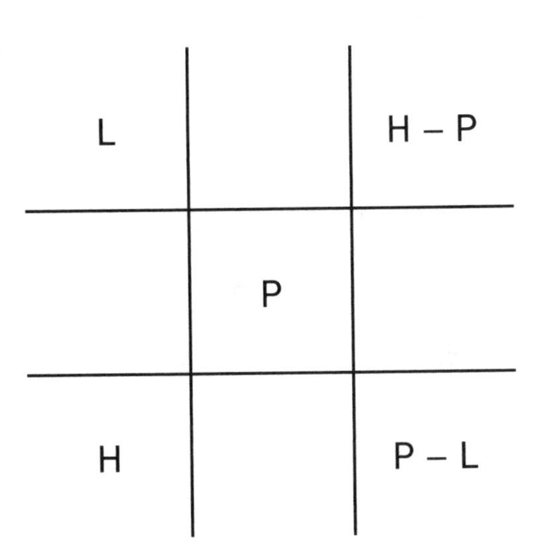

Where L is the lower strength (or concentration), H is the higher strength (or concentration), P is the strength or concentration of the Product, H – P is the relative amount of L needed for the Product and P – L is the relative amount of H needed for the Product.

POWDER VOLUME

The final volume of the constituted product FV equals the volume of the diluent D plus the powder volume PV.

$$FV = D + PV$$

PEDIATRIC DOSES

Because of the many variables, conversion formulas for pediatric doses are rarely used in the pharmacy. Doses are generally given by the physician. **Children's doses are stated by kg of body weight (dose/kg).** Since 1 kg = 2.2 lb, you can solve for the prescribed dose by using a proportion equation if you know the child's body weight. See the following example.

An antibiotic IV is prescribed for an infant. The dose is to be 15 mg/kg twice a day. The baby weighs 12 lbs. How much drug is to be given for one dose? First the infant's weight in kilograms should be calculated.

$$x \text{ kg} / 12 \text{ lb} = 1 \text{ kg} / 2.2 \text{ lb}$$

$$x \text{ kg} = 12 \cancel{\text{lb}} \text{ times } \frac{1 \text{ kg}}{2.2 \cancel{\text{lb}}} = \frac{12 \text{ kg}}{2.2} = 5.45 \text{ kg}$$

The next part of this problem can be solved with a simple equation.

$$\text{one dose} = 15 \text{ mg times } 5.45 = 81.75 \text{ mg}$$

IN THE WORKPLACE ACTIVITIES

1. Calculate the number of tablets needed if the patient must take four tablets every four hours for five days, followed by three tablets every six hours for five days, followed by one tablet every 12 hours for 25 days, and then one tablet every other day for six days.
2. Calculate the total amount of suspension that will be needed if a patient must take 10 mL every eight hours for five days, followed by 10 mL every 12 hours for three days, followed by 5 mL every 12 hours for 10 days.

TOTAL PARENTERAL NUTRITION

A TPN order calls for the amounts on the left (including additives) to be made from the items on the right. The total volume is to be 1,000 mL. How much of each ingredient and how much sterile water do you need to prepare this TPN?

TPN Order		**On Hand**	
Aminosyn® 4.25%		Aminosyn® 8.5%	1,000 mL
dextrose 20%		dextrose 50%	500 mL
Additives:			
KCl	24 mEq	KCl 2 mEq/ml	20 mL
MVI	5 ml	MVI	10 mL
NaCl	24 mEq	NaCl 4.4 mEq/ml	20 mL

Figure out the amounts and enter the answer on the blank line.

1. Aminosyn® ____________________

2. dextrose ____________________

3. KCl ____________________

4. MVI ____________________

5. NaCl ____________________

6. sterile water ____________________

PROBLEMS

1. You have a prescription that calls for 1 cap po tid x 10 days. How many capsules are needed?

2. You have a prescription that calls for 1 cap po tid x 7 days. How many capsules are needed?

3. If a compounding order calls for Flagyl® 125 mg bid x 7 days, and only 500 mg tablets are available, how many tablets will it take to fill the order?

4. How much talc is needed for an order for 30 g of the following compound: Nupercainal® ointment 4%, zinc oxide 20%, talc 2%?

5. A prescription calls for 200 mg of a drug that you have in a 10 mg/15 mL concentration. How many mL of the liquid do you need?

6. An IV requires the addition of 40 mEq potassium chloride (KCl). You have a vial of KCl at a concentration of 20 mEq per 10 mL. How many mL should be added?

7. If 100 grams of dextrose is ordered using a 50% dextrose solution, how many mL are needed?

8. A prescription calls for 0.36 mg of a drug that you have in 50 mcg/mL concentration. How many mL do you need?

9. The infusion rate of an IV is 300 mL over 5 hours. What is the mL/minute rate?

10. The infusion rate of an IV is 1,000 mL over 12 hours. What is the rate in mL per minute?

11. An IV order calls for administration of 1.5 mL/minute of a solution for two hours. How much solution will be needed.

12. If a physician orders 25% dextrose 1,000 mL and all you have is 70% dextrose 1,000 mL, how much 70% dextrose and how much sterile water will be used?

13. If a physician orders 20% dextrose 1,000 mL and all you have is 70% dextrose 1,000 mL, how much 70% dextrose and how much sterile water will be used?

14. If a physician orders 25% dextrose 500 mL and you have 50% dextrose 1,000 mL, how much 50% dextrose and how much sterile water do you need?

15. What is the powder volume if 27 mL of water are in added to reconstitute a liquid antibiotic and the final volume is 60 mL

STUDY NOTES

Use this area to write important points you'd like to remember.

— 7 —

PRESCRIPTIONS

KEY CONCEPTS

Test your knowledge by covering the information in the right-hand column.

prescription	An instruction from a medical practitioner that authorizes a patient to be issued a drug or device.
computer processes	The computer system evaluates the drug against stored information and processes third-party billing.
pharmacy abbreviations	Pharmacy abbreviations are often based on Latin words and use Roman numerals for numbers.
Signa	The directions for use.
prescription information safety checklist	A safety checklist can be used to help ensure prescriptions are filled safely and accurately.
fraud, waste, or abuse (FWA)	Pharmacy technicians play a vital role in the effort to prevent, detect, and report possible fraud, waste, or abuse of prescription drug benefits.
the fill process	Once prescription information is finalized in the computerized prescription system, a label and receipt are printed, and the pharmacy technician completes the fill process by placing the correct amount of medication into an appropriate container and labeling it correctly.
pharmacist check	If a prescription has been prepared by a technician, there is a final check by the pharmacist to make sure that it is correct.
judgment questions	Technicians must request the advice of the pharmacist whenever judgment is required.
safety considerations	A checklist can be used to ensure prescriptions are filled accurately and safely.
high-alert medications	Medications that are known to cause significant harm if an error is made.

tall man lettering Uppercase letters used within a drug name to highlight differences between look-alike drug names.

days supply The number of days the prescribed quantity of medication will last when taken as directed.

transfers A prescription may be transferred from one pharmacy to another in accordance with state laws and rules.

label The general purpose of the prescription label is to provide information to the patient regarding the dispensed medication and how to take it. Additionally, the label includes information about the pharmacy, the patient, the prescriber, and the prescription or transaction number assigned to the prescription. Computer-generated prescription labels must be placed on containers so they are easy to locate and easy to read.

auxiliary labels Many computerized prescription systems will automatically indicate which auxiliary labels to use with each drug.

medication orders Used in institutional settings instead of a prescription form.

STUDY NOTES

Use this area to write important points you'd like to remember.

COMMON PHARMACY ABBREVIATIONS

Here are the most common pharmacy abbreviations.

ROUTE

ad	right ear
as, al	left ear
au	each ear
od	right eye
os, ol	left eye
ou	each eye
po	by mouth
SL	sublingually, under the tongue
top	topically, locally
pr	rectally, into the rectum
pv	vaginally, into the vagina
inh	inhalation, inhale
per neb	by nebulizer
SC, subc, subq	subcutaneously
im, IM	intramuscular
iv, IV	intravenous
ivp, IVP	intravenous push
IVPB	intravenous piggyback

FORM

tab	tablet
caps	capsules
SR, XR, XL	slow/extended release
sol	solution
susp	suspension
syr	syrup
liq	liquid
supp	suppository
cm	cream
ung, oint	ointment

TIME

bid	twice a day
tid	three times a day
qid	four times a day
am/q am	morning/each morning
pm	afternoon or evening
hs	at bedtime
prn	as needed
ac	before food, before meals
pc	after food, after meals
stat	immediately
q_h	every hour

MEASUREMENT

I, II	one, two, etc.
ss	one-half
gtt	drop
mL	milliliter/millilitre
tsp	teaspoon
tbsp	tablespoon
fl oz	fluid ounce
L	liter/litre
mcg, µg	microgram
mg	milligram
g, G, gm	gram
mEq	milliequivalent
aa	of each
ad	to, up to
aq ad	add water up to
qs, qs ad	add sufficient quantity to make

OTHER

UD	as directed
NR, ø	no refill
DAW	dispense as written
c̄, w	with
s̄, w/o	wihout

Note that the use of periods in abbreviations varies greatly. It is important to be able to recognize abbreviations with or without periods.

FILL IN THE BLANK

Answers are at the end of the book.

Abbreviation	No.	Answer
aa	1.	________
ac	2.	________
ad	3.	________
al	4.	________
aq ad	5.	________
as	6.	________
au	7.	________
bid	8.	________
$\bar{c}$	9.	________
caps	10.	________
cm	11.	________
DAW	12.	________
fl oz	13.	________
g, G, gm	14.	________
gtt	15.	________
hs	16.	________
I	17.	________
II	18.	________
im, IM	19.	________
inh	20.	________
iv, IV	21.	________
ivp, IVP	22.	________
IVPB	23.	________
L	24.	________
liq	25.	________
mcg	26.	________
mEq	27.	________
mg	28.	________
mL	29.	________
NR, ø	30.	________
od	31.	________
ol	32.	________
os	33.	________
ou	34.	________
pc	35.	________
per neb	36.	________
pm	37.	________
po	38.	________
pr	39.	________
prn	40.	________
pv	41.	________
q am	42.	________
q_h	43.	________
qid	44.	________
qs ad	45.	________
$\bar{s}$	46.	________
SC	47.	________
SR	48.	________
SL	49.	________
sol	50.	________
ss	51.	________
stat	52.	________
supp	53.	________
susp	54.	________
syr	55.	________
tid	56.	________
tab	57.	________
tbsp	58.	________
top	59.	________
tsp	60.	________
ung	61.	________
UD	62.	________
w	63	________
w/o	64.	________

THE PRESCRIPTION

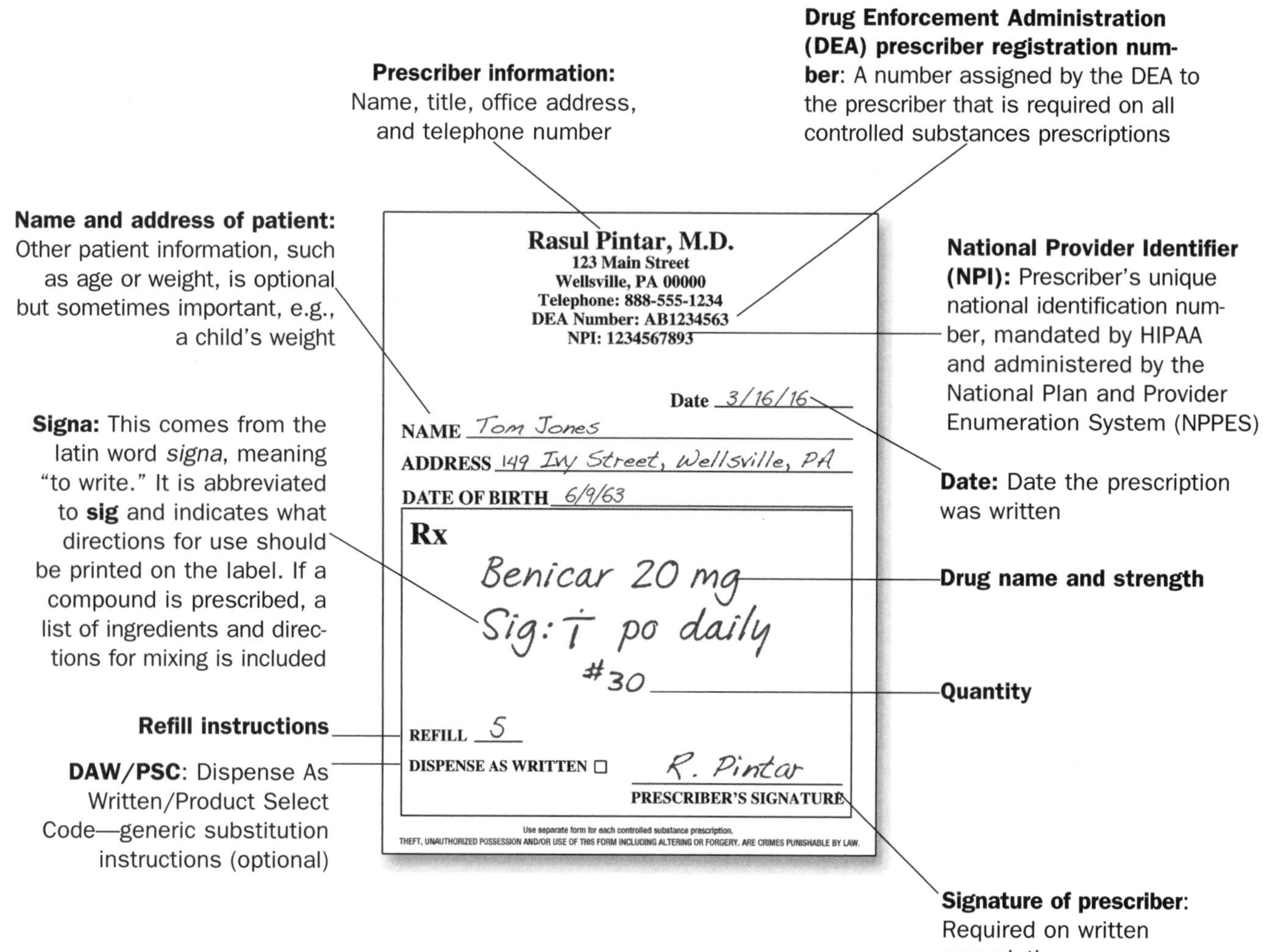

PRESCRIPTION LABELS

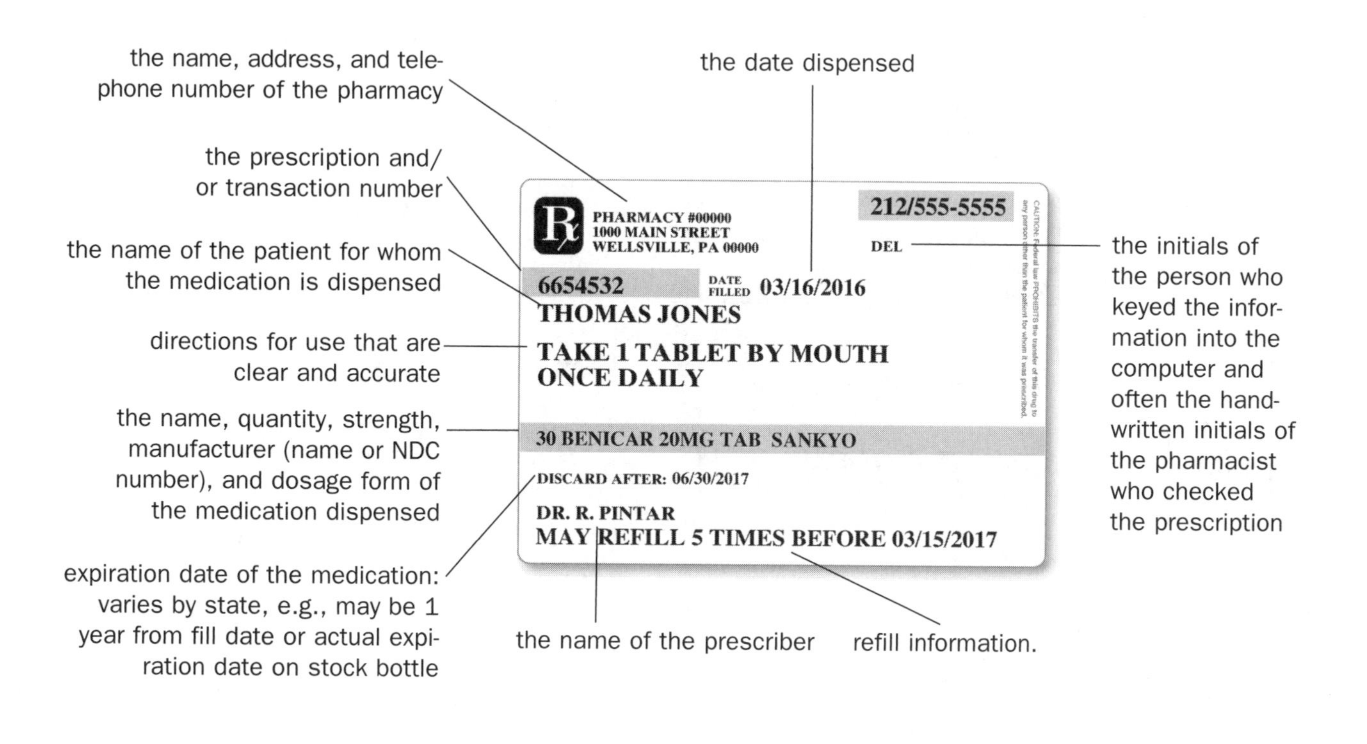

AUXILIARY LABELS

Colored auxiliary labels may be applied to the prescription container to provide additional information to the patient (e.g., Shake Well, Keep Refrigerated, Take With Food or Milk). Many computerized prescription systems will automatically print out the appropriate labels to use.

For many drugs, the timing of the dose is almost as important as taking the dose. For example, taking with food or milk enhances the absorption of some drugs but greatly diminishes the effectiveness of others.

Prescriptions for controlled substances from Schedules II, III, and IV must carry the following warning, which is preprinted on the label: *Caution: Federal law prohibits the transfer of this drug to any person other than the patient for whom it was prescribed.*

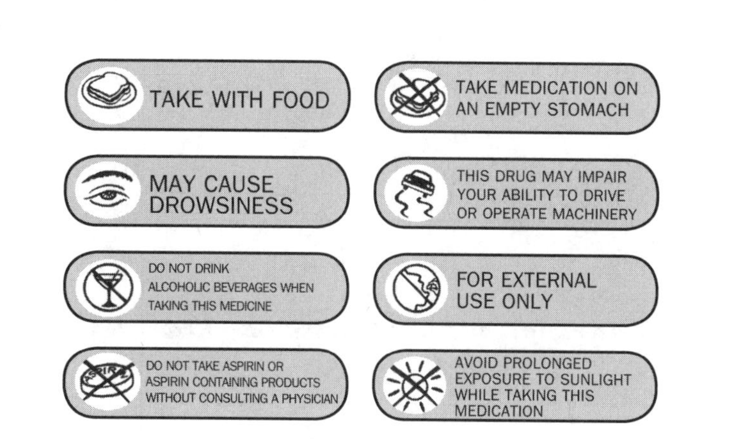

PRACTICE PRESCRIPTIONS

Identify the elements on these prescriptions by answering in the space beneath the question.

Dr. A.B. Cain
123 Main Street
Wellsville, PA 00000
Telephone: 888/555-1234
DEA number AB1234563
NPI: 3333333334

Date March 1, 2016

NAME D.H. Doe

ADDRESS 345 Maple St, Wellsville PA
D.O.B - 2/23/54

Rx

Prozac 20 mg
Sig: i cap p.o. daily
#30

REFILL X2

DISPENSE AS WRITTEN ☑ A.B. Cain

PRESCRIBER'S SIGNATURE

Use separate form for each controlled substance prescription.
THEFT, UNAUTHORIZED POSSESSION AND/OR USE OF THIS FORM INCLUDING ALTERATIONS OR FORGERY, ARE CRIMES PUNISHABLE BY LAW.

1. What is the name of the drug?
2. What is the strength?
3. What is the dosage form?
4. What is the route of administration?
5. What is the dosage?
6. How many refills are there?
7. Can there be generic substitution?

JANE T. DOE, MD
1002 Main Street
WELLSVILLE, PA, 00000
(212) 555-5555
NPI: 2222222228

Date 6/5/2016

NAME John Jones

ADDRESS

Rx

Triamcinolone 0.1% Cream 15g
Apply BID

REFILL 1

Doe

DEA No.

PRESCRIBER'S SIGNATURE

Use separate form for each controlled substance prescription.
THEFT, UNAUTHORIZED POSSESSION AND/OR USE OF THIS FORM INCLUDING ALTERATIONS OR FORGERY, ARE CRIMES PUNISHABLE BY LAW.

1. What is the strength of triamcinolone cream ordered by Dr. Doe?
2. How often should the medication be applied?

PRACTICE PRESCRIPTIONS

Identify the elements on these prescriptions by answering in the space beneath the question.

JANE T. DOE, MD
1002 Main Street
WELLSVILLE, PA, 00000
(212) 555-5555
NPI: 2222222228

Date 6/5/2016

NAME Andrew Jones

ADDRESS

Rx

Protonix 40 mg #60
T BID x 8 weeks

REFILL 1

DEA No.

Doe
PRESCRIBER'S SIGNATURE

Use separate form for each controlled substance perscription
THEFT, UNAUTHORIZED POSSESSION AND/OR USE OF THIS FORM INCLUDING ALTERATIONS OR FORGERY, ARE CRIMES PUNISHABLE BY LAW.

1. This prescription is written for Protonix®. If the medication is taken as prescribed, every day, how many days will this prescription last?

2. How many refills have been ordered for this prescription?

JANE T. DOE, MD
1002 Main Street
WELLSVILLE, PA, 00000
(212) 555-5555
NPI: 2222222228

Date 6/5/2016

NAME Steve Jones

ADDRESS

Rx

Atrovent HFA #1 2-3 puff TID
Flovent HFA 110 mcg #1 2 puff BID

REFILL 5

DEA No.

Doe
PRESCRIBER'S SIGNATURE

Use separate form for each controlled substance perscription
THEFT, UNAUTHORIZED POSSESSION AND/OR USE OF THIS FORM INCLUDING ALTERATIONS OR FORGERY, ARE CRIMES PUNISHABLE BY LAW.

1. By reading the box for the Atrovent® HFA you find that each inhaler contains 200 inhalations. What is the days supply that should be entered in the computer for the Atrovent prescription?

2. Flovent® HFA 110 mcg is dispensed in a 12 g canister that contains 120 metered doses. What is the days supply that should be entered in the computer for the Flovent® prescription?

PRACTICE PRESCRIPTIONS

Identify the elements on these prescriptions by answering in the space beneath the question.

JANE T. DOE, MD
1002 Main Street
WELLSVILLE, PA, 00000
(212) 555-5555
NPI: 2222222228

Date 6/5/2016

NAME Samuel Jones

ADDRESS

Rx

Synthroid 0.05 mg DAW #30
i po daily

REFILL 5

DEA No.

Doe
PRESCRIBER'S SIGNATURE

Use separate form for each controlled substance perscription
THEFT, UNAUTHORIZED POSSESSION AND/OR USE OF THIS FORM INCLUDIND ALTERATIONS OR FORGERY, ARE CRIMES PUNISHABLE BY LAW.

1. This patient has a dual co-pay of $15 for brand and $5 for generic. The patient has requested the generic for this prescription. What is written on the prescription that does not allow the generic to be dispensed?

JANE T. DOE, MD
1002 Main Street
WELLSVILLE, PA, 00000
(212) 555-5555
NPI: 2222222228

Date 6/5/2016

NAME Francis Jones

ADDRESS

Rx

Miacalcin 200 IU /nasal
i inh daily (alt. nostrils)

REFILL 1

DEA No.

Doe
PRESCRIBER'S SIGNATURE

Use separate form for each controlled substance perscription
THEFT, UNAUTHORIZED POSSESSION AND/OR USE OF THIS FORM INCLUDIND ALTERATIONS OR FORGERY, ARE CRIMES PUNISHABLE BY LAW.

1. How should the patient use this medication?

PRACTICE PRESCRIPTIONS

Identify the elements on these prescriptions by answering in the space beneath the question.

JANE T. DOE, MD
1002 Main Street
WELLSVILLE, PA, 00000
(212) 555-5555
NPI: 2222222228

Date 6/5/2016

NAME Sandra Jones

ADDRESS

Rx

Metrogel Vaginal 70 g
ī applic PV qhs x 5D

REFILL nr

DEA No.

Doe
PRESCRIBER'S SIGNATURE

Use separate form for each controlled substance perscription
THEFT, UNAUTHORIZED POSSESSION AND/OR USE OF THIS FORM INCLUDIND ALTERATIONS OR FORGERY, ARE CRIMES PUNISHABLE BY LAW.

1. What does 70 g mean for this prescription?

2. What does PV mean?

JANE T. DOE, MD
1002 Main Street
WELLSVILLE, PA, 00000
(212) 555-5555
NPI: 2222222228

Date 6/5/2016

NAME Joyce Jones

ADDRESS

Rx
Premarin 1.25 mg #21
ī daily x 21, off 7

Provera 10 mg #5
1 daily days 10-14

REFILL 5

DEA No.

Doe
PRESCRIBER'S SIGNATURE

Use separate form for each controlled substance perscription
THEFT, UNAUTHORIZED POSSESSION AND/OR USE OF THIS FORM INCLUDIND ALTERATIONS OR FORGERY, ARE CRIMES PUNISHABLE BY LAW.

1. How many days will the Premarin® prescription last?

2. How many days will the Provera® prescription last?

PRACTICE PRESCRIPTIONS

Identify the elements on these prescriptions by answering in the space beneath the question.

JANE T. DOE, MD
1002 Main Street
WELLSVILLE, PA, 00000
(212) 555-5555
NPI: 2222222228

Date 6/5/2016

NAME Cindy Jones

ADDRESS

Rx

Ortho-Novum 777 28 day
T daily

REFILL 6

DEA No.

Doe

PRESCRIBER'S SIGNATURE

Use separate form for each controlled substance perscription
THEFT, UNAUTHORIZED POSSESSION AND/OR USE OF THIS FORM INCLUDIND ALTERATIONS OR FORGERY, ARE CRIMES PUNISHABLE BY LAW.

1. What is the total number of compacts indicated by this prescription (including the original fill plus refills)?

JANE T. DOE, MD
1002 Main Street
WELLSVILLE, PA, 00000
(212) 555-5555
NPI: 2222222228

Date 6/5/2016

NAME Jane Jones

ADDRESS

Rx

Bactrim DS 20
T po BID

REFILL nr

DEA No.

Doe

PRESCRIBER'S SIGNATURE

Use separate form for each controlled substance perscription
THEFT, UNAUTHORIZED POSSESSION AND/OR USE OF THIS FORM INCLUDIND ALTERATIONS OR FORGERY, ARE CRIMES PUNISHABLE BY LAW.

1. As you are entering this prescription in the computer, an allergy warning is displayed on the computer screen. What type of allergy does Jane Jones have?

PRACTICE PRESCRIPTIONS

Identify the elements on these prescriptions by answering in the space beneath the question.

JANE T. DOE, MD
1002 Main Street
WELLSVILLE, PA, 00000
(212) 555-5555
NPI: 2222222228

Date 6/5/2016

NAME Ann Jones

ADDRESS

Rx

Cefadroxil 500mg/5ml
i ʒ po BID x 10

REFILL nr

DEA No.

Doe
PRESCRIBER'S SIGNATURE

Use separate form for each controlled substance perscription
THEFT, UNAUTHORIZED POSSESSION AND/OR USE OF THIS FORM INCLUDIND ALTERATIONS OR FORGERY, ARE CRIMES PUNISHABLE BY LAW.

1. What directions should be placed on the label for this prescription?

JANE T. DOE, MD
1002 Main Street
WELLSVILLE, PA, 00000
(212) 555-5555
NPI: 2222222228

Date 6/5/2016

NAME Tom Jones

ADDRESS

Rx

Amoxicillin 500 mg 2 po BID x 7
Biaxin 500 mg po BID x 7
Aciphex 20 mg po BID x 7

REFILL nr

DEA No.

Doe
PRESCRIBER'S SIGNATURE

Use separate form for each controlled substance perscription
THEFT, UNAUTHORIZED POSSESSION AND/OR USE OF THIS FORM INCLUDIND ALTERATIONS OR FORGERY, ARE CRIMES PUNISHABLE BY LAW.

1. How many capsules should be dispensed for the amoxicillin?

2. How many tablets should be dispensed for the Biaxin®?

3. How many tablets should be dispensed for the Aciphex®?

PRACTICE MEDICATION ORDERS

Identify the elements on this medication order by answering in the space beneath the question.

DOCTOR'S ORDERS

PATIENT IDENTIFICATION

099999999 675-01
SMITH, JOHN
12/06/1950
DR. P. JOHNSON

DATE	TIME	DOCTOR'S ORDERS 1	DATE/TIME INITIALS	DATE/TIME INITIALS
1/31/16	2200	Admit patient to 6th floor		
		Pneumonia, Dehydration		
		All: PCN- Rash		
		Order CBC, chem-7, blood cultures stat		
		NS @ 125ml/hr IV		
		Dr Johnson x2222		

DATE	TIME	DOCTOR'S ORDERS 2	DATE/TIME INITIALS	DATE/TIME INITIALS
2/01/16	300	Tylenol 650mg po q4-6 hrs PRN for Temp>38°C		
		Verbal Order Dr Johnson/ Jane Doe, RN		

DATE	TIME	DOCTOR'S ORDERS 3	DATE/TIME INITIALS	DATE/TIME INITIALS
2/01/16	600	Start Clarithromycin 500mg po q12h		
		Multivitamin po daily		
		Order CXR for this a.m.		
		Dr Johnson x2222		

1. What is the patient's disorder/condition?

2. Does the patient have allergies?

3. What route and dosage is ordered for the Tylenol® 650 mg?

4. What orders did the physician sign for?

5. What route and dosage are prescribed for the Clarithromycin?

6. What route and dosage are prescribed for the multivitamin?

PRACTICE MEDICATION ORDERS

Identify the elements on this medication order by answering in the space beneath the question.

PATIENT: John Smith
AGE: 35
SEX: m
CHART #: #123555

CITY HOSPITAL
PHYSICIAN'S ORDERS

ALLERGIES:

DIAGNOSIS:

DATE	TIME	ORDERS	SIGNATURE	COMPLETED OR DISCONTINUED NAME	DATE	TIME
		Meds:				
6/5/16	12:34 pm	① Lopressor 50 mg po daily				
		② HCTZ 25 mg po daily				
		③ Sonata 5 mg po hs prn				
		Doe				

PHARMACY COPY

1. What three medications have been ordered?

2. How often should the Sonata® be taken?

PRACTICE MEDICATION ORDERS

Identify the elements on these medication orders by answering in the space beneath the question.

PATIENT: Mary Smith
AGE: 35
SEX: F
CHART #: #123505

CITY HOSPITAL
PHYSICIAN'S ORDERS

ALLERGIES:

DIAGNOSIS:

COMPLETED OR DISCONTINUED

DATE	TIME	ORDERS	SIGNATURE	NAME	DATE	TIME
6/5/16	12:34pm	Docusate Sod 100 mg po BID today				
		Start Metamucil tomorrow				
		i tsp BID				

PHARMACY COPY

1. What medication has been ordered for today?

2. What medication should be started tomorrow?

PATIENT: Andrew Smith
AGE: 45
SEX: M
CHART #: 123616

CITY HOSPITAL
PHYSICIAN'S ORDERS

ALLERGIES:

DIAGNOSIS:

COMPLETED OR DISCONTINUED

DATE	TIME	ORDERS	SIGNATURE	NAME	DATE	TIME
6/5/16	10:22am	Vit B12 1000 MCG IM Stat				
		Multivitamin i po daily				

PHARMACY COPY

1. What does STAT mean in the order for vitamin B-12?

2. What is the route of administration for vitamin B-12?

PRACTICE MEDICATION ORDERS

Identify the elements on these medication orders by answering in the space beneath the question.

PATIENT: Steve Smith
AGE: 45
SEX: M
CHART #: 123777

CITY HOSPITAL
PHYSICIAN'S ORDERS

ALLERGIES: Penicillin

DIAGNOSIS:

DATE	TIME	ORDERS	SIGNATURE	COMPLETED OR DISCONTINUED NAME	DATE	TIME
6/5/16	10:15am	IVF NS @ 100 cc/hr for 2 days				
		HCTZ 25 mg po daily				
		Diovan 80 mg po daily				
			Doe			

PHARMACY COPY

1. What is the rate for the IV in this medication order?

2. What medications are to be given orally?

PATIENT: Barbara Smith
AGE: 45
SEX: F
CHART #: 123718

CITY HOSPITAL
PHYSICIAN'S ORDERS

ALLERGIES: Codeine

DIAGNOSIS:

DATE	TIME	ORDERS	SIGNATURE	COMPLETED OR DISCONTINUED NAME	DATE	TIME
6/5/16	10:10am	FS AC + HS				
		Glyburide 5mg po daily				
		Ambien 5mg po hs prn				
			Doe			

PHARMACY COPY

1. What medications have been ordered?

2. Which medication would be administered at bedtime, as needed?

FILL IN THE KEY TERM

Use these key terms to fill in the correct blank. Answers are at the end of the book.

auxiliary label
DAW
DEA number
DUR
formulary
extemporaneous compounding
high-alert medications
look-alikes
medication order
NPI
product select codes
refills
Rx
transfers

1. ____________________ : Much like filling an original prescription except that information has already been entered into the computer.
2. ____________________ : The pharmaceutical preparation of a medication from ingredients.
3. ____________________ : A list of drugs approved by the insurance provider or a formulary committee.
4. ____________________ : Generic substitution instructions.
5. ____________________ : The additional warning labels that are placed on filled prescription containers.
6. ____________________ : The form used to prescribe medications for patients in institutional settings.
7. ____________________ : Drug names that have similar appearance, particularly when written.
8. ____________________ : Abbreviation for the Latin word "recipe."
9. ____________________ : Dispense As Written, meaning generic substitution not allowed.
10. ____________________ : Medications that are known to cause significant harm if an error is made.
11. ____________________ : A prescription may be transferred from one pharmacy to another in accordance with state laws and rules.
12. ____________________ : Drug utilization review.
13. ____________________ : Prescriber's unique national identification number.
14. ____________________ : Required on all controlled substance prescriptions.

TRUE/FALSE

Indicate whether the statement is true or false in the blank. Answers are at the end of the book.

_____ 1. In addition to the primary prescribers, nurse practitioners, pharmacists, and physician assistants are allowed to write prescriptions in some states.

_____ 2. In hospitals, both the pharmacy and nursing unit have a copy of the medication order.

_____ 3. Rules and regulations for dispensing prescriptions for hospital inpatients are the same as for community pharmacies.

_____ 4. In institutional settings, nursing staff generally administer medications to patients.

_____ 5. Pharmacy technicians do not need to look for signs of fraud, waste, or abuse.

_____ 6. Counseling patients on the use of OTC medications should be done by the pharmacist.

_____ 7. One of the primary purposes of the prescription label is to provide the patient with clear instructions on how to take the medication.

_____ 8. Unit dose packaging is never used in institutional pharmacy.

_____ 9. Label directions always start with a noun.

_____ 10. A DEA number is required on all prescriptions.

EXPLAIN WHY

Explain why these statements are true or important. Check your answers in the text. Discuss any questions you may have with your instructor.

1. Why must the pharmacist always check the filled prescription before it is dispensed to the patient?
2. Why must the directions for use be clear and understandable to the patient in the community setting?
3. What are the differences between the prescription and the medication order? Why?
4. Why must prescriptions be written in ink?
5. Why are auxiliary labels important?

IN THE WORKPLACE

Use this tool to practice and check your skills in gathering prescription information.

Pharmacy Technician Checklist
PRESCRIPTION INFORMATION

	YES	NO
Is the patient's full name clearly written on the prescription?	______	______
Has a nickname or initial been used?	______	______
Is the patient's date of birth, street address, telephone number, insurance information, and allergy information already on file?	______	______
Is the medication for an over-the-counter product that the patient can receive without a prescription?	______	______
Is the prescription for a Schedule II drug?	______	______
When was the prescription written?	______	______
How many days or weeks has it been since it was written?	______	______
Is the drug available in the quantity written?	______	______
Does it require compounding?	______	______
Is the prescription suspicious in any way?	______	______
Is it written on a legitimate prescription blank and all in the same hand-writing and with the same ink?	______	______
Are there any signs of alteration of quantities, strength, or the name of the drug?	______	______
Is this a drug with the potential for abuse and if so do the quantities and directions seem appropriate?	______	______
Is it signed by the prescriber?	______	______
Are there any other signs of fraud, waste, or abuse (FWA), such as the use of another person's insurance card or repeated early refills?	______	______

IN THE WORKPLACE

Use this tool to practice and check your safety skills during the fill process.

Pharmacy Technician Checklist
FILL PROCESS

	YES	NO
Is the prescription for a high-alert medication?	______	______
Are the instructions clear and logical?	______	______
Are there leading or trailing zeroes?	______	______
Are the label directions clear?	______	______
Is the route of administration correct?	______	______
Are you selecting the correct drug to fill the prescription?	______	______
Are there any look-alike drug names that could be confused with the intended medication?	______	______
Is the prescription for tablets or capsules? Is it for extended or sustained release?	______	______
Are you dispensing the right prescription to the right person?	______	______
Should the pharmacist be called to give patient counseling?	______	______

IN THE WORKPLACE ACTIVITY

1. Prepare labels for the prescriptions on pages 70–75 of this book and appropriately place them on prescription vials, and identify which auxiliary labels should be used.

CHOOSE THE BEST ANSWER

Answers are at the end of the book.

1. When a hospital patient has been taking a drug at home that is not on the hospital's formulary, the ________________ plays a key role in helping the physician identify the most appropriate equivalent for that drug.
   a. Nurse
   b. Therapist
   c. Pharmacist
   d. Pharmacy technician

2. An alert or message generated by the pharmacy or insurance company computer system notifying the pharmacist of a potential drug safety concern or payment concern.
   a. DAW
   b. DUR
   c. OBRA
   d. FWA

3. The _______ uses abbreviations to communicate essential information such as the dosage form, dosage regimen, and route of administration.
   a. Signa
   b. NPI
   c. PSC
   d. FWA

4. When a prescription is written for a medication that is not commercially available, the medication can be prepared by mixing the ingredients required and this is called
   a. alligation.
   b. extemporaneous compounding.
   c. trituration.
   d. admixture.

5. If a new prescription has been prepared by a pharmacy technician, the final check is done by the
   a. lead pharmacy technician.
   b. pharmacist.
   c. patient.
   d. physician.

6. An example of an over-the-counter medication that was previously available by prescription only:
   a. Nexium
   b. Lasix
   c. Lanoxin
   d. OxyContin

7. The DEA Number is required on prescriptions for
   a. legend drugs.
   b. medical devices.
   c. controlled substances.
   d. OTCs.

8. An example of a high-alert medication:
   a. Lasix
   b. Lanoxin
   c. Coumadin
   d. OxyContin

9. The NPI was mandated by
   a. the APhA.
   b. the ASHP.
   c. HIPAA.
   d. the ISMP.

10. The number of identity checks that are recommended to determine if you are dispensing the right prescription to the right patient.
   a. 0
   b. 1
   c. 2
   d. 3

11. When the Sig contains t.i.d., the medication should be taken __________ times a day.
   a. one
   b. two
   c. three
   d. four

12. Antibiotic suspensions that are reconstituted in-house are typically stable for _____ days.
   a. 10 to 14
   b. 14 to 21
   c. 21 to 28
   d. 30 to 60

13. A package containing a single dose of a medication.
   a. Extemporaneous
   b. Unit dose
   c. Formulary
   d. FWA

14. Directions should start with a
   a. verb.
   b. noun.
   c. adjective.
   d. pronoun.

15. For the prescription: Bactrim® DS #20 $\dot{\bar{\text{i}}}$ b.i.d. NR, how many days should the prescription last?
   a. 2
   b. 5
   c. 10
   d. 20

STUDY NOTES

Use this area to write important points you'd like to remember.

— 8 —

ROUTES & FORMULATIONS

KEY CONCEPTS

Test your knowledge by covering the information in the right-hand column.

formulations	Drugs are contained in dosage units called formulations, or dosage forms. There are many dosage forms and many different routes to administer them.
route of administration	Routes are classified as enteral or parenteral. Enteral refers to anything involving the tract from the mouth to the rectum. There are three enteral routes: oral, sublingual, and rectal. Any route other than oral, sublingual, and rectal is considered a parenteral administration route. Oral administration is the most frequently used route of administration.
local and systemic effects	A local effect occurs when the drug activity is at the site of administration (e.g., eyes, ears, nose, skin). A systemic effect occurs when the drug is introduced into the circulatory system.
oral administration	The stomach has a pH around 1–2. Certain drugs cannot be taken orally because they are degraded or destroyed by stomach acid and intestinal enzymes. Drugs administered by liquid dosage forms generally reach the circulatory system faster than drugs formulated in solid dosage forms.
inactive ingredients	Various ingredients contained in oral formulations besides the active drug, including binders, lubricants, fillers, diluents, and disintegrants.
gastrointestinal action	The disintegration and dissolution of tablets, capsules, and powders generally begins in the stomach, but will continue to occur when the stomach empties into the intestine. Modified release formulations extend dissolution over a period of hours and provide a longer duration of effect compared to plain tablets. Enteric coated tablets prevent the tablet from disintegrating until it reaches the higher pHs of the intestine.
modified release formulation	Special oral formulations that release the drug with a longer duration of action compared to a conventional tablet, capsule, or powder.

sublingual administration

These tablets are placed under the tongue. They are generally fast-dissolving, uncoated tablets that contain highly water-soluble drugs. When the drug is released from the tablet, it is quickly absorbed into the circulatory system since the membranes lining the mouth are very thin and there is a rich blood supply to the mouth.

rectal administration

Rectal administration may be used to achieve a variety of systemic effects, including: asthma control, antinausea, antimotion sickness, and anti-infective. However, absorption from rectal administration is erratic and unpredictable. The most common rectal administration forms are suppositories, solutions, and ointments.

parenteral administration

Parenteral routes are often preferred when oral administration causes drug degradation or when a rapid drug response is desired, as in an emergency situation. The parenteral routes requiring a needle are intravenous, intramuscular, intradermal, and subcutaneous. These solutions must be sterile (bacteria-free), have an appropriate pH, and be limited in volume.

intravenous formulations

Intravenous dosage forms are administered directly into a vein (and the blood supply). Most solutions are aqueous (water based), but they may also have glycols, alcohols, or other nonaqueous solvents in them.

infusion

Infusion is the gradual intravenous injection of a volume of fluid into a patient.

intravenous sites

Several sites on the body are used to intravenously administer drugs: the veins of the antecubital area (in front of the elbow), the back of the hand, and some of the larger veins in the foot. On some occasions, a vein must be exposed by a surgical cut.

intramuscular injections

The principal sites of injection are the gluteal maximus (buttocks), deltoid (upper arm), and vastus lateralis (thigh) muscles. Intramuscular injections generally result in lower but longer lasting blood concentrations than with intravenous administration.

subcutaneous injections

Injection sites include the back of the upper arm, the front of the thigh, the lower portion of the abdomen and the upper back. The subcutaneous (SC, SQ) route can be used for both short-term and very long-term therapies. Insulin is the most important drug routinely administered by this route.

intradermal injections

Intradermal injections involve small volumes that are injected into the top layer of skin. The usual site for intradermal injections is the anterior surface of the forearm.

ophthalmic formulations

Every ophthalmic product must be manufactured to be sterile in its final container. A major problem of ophthalmic administration is the immediate loss of a dose by natural spillage from the eye.

KEY CONCEPTS (CONT'D)

Test your knowledge by covering the information in the right hand column.

intranasal formulations Intranasal formulations are primarily used for their decongestant activity on the nasal mucosa, the cellular lining of the nose. The drugs that are typically used are decongestants, antihistamines, and corticosteroids. Nasal administration often causes amounts of the drug to be swallowed, in some cases this may lead to a systemic effect.

inhalation formulations Inhalation dosage forms are intended to deliver drugs to the pulmonary system (lungs). Most of the inhalation dosage forms are aerosols that depend on the power of compressed or liquefied gas to expel the drug from the container. Gaseous or volatile anesthetics are the most important drugs administered via this route. Other drugs administered affect lung function, act as bronchodilators, or treat allergic symptoms. Examples of drugs administered by this route are adrenocorticoid steroids (beclomethasone), bronchodilators (epinephrine, isoproterenol, metaproterenol, albuterol), and antiallergics (cromolyn sodium).

dermal formulations Most dermal dosage forms are used for local (topical) effects on or within the skin. Dermal formulations are used to treat minor skin infections, itching, burns, diaper rash, insect stings and bites, athlete's foot, corns, calluses, warts, dandruff, acne, psoriasis, and eczema. The major disadvantage of this route of administration is that the amount of drug that can be absorbed will be limited to about 2 mg/hour.

vaginal administration Formulations for this route of administration are: solutions, powders for solutions, ointments, creams, aerosol foams, suppositories, tablets, and IUDs.

EXPLAIN WHY

Explain why these statements are true or important. Check your answers in the text. Discuss any questions you may have with your instructor.

1. Give three reasons why a drug might not be used for oral administration.
2. Give three reasons why a drug might not be used for parenteral administration.
3. Why do most parenterals require skilled personnel to administer them?
4. Why is the pH of intravenous solutions important?
5. Why is the development of infusion pumps important?
6. Why is a spacer sometimes used with a metered dose inhaler?

TRUE/FALSE

Indicate whether the statement is true or false in the blank. Answers are at the end of the book.

_____ 1. Buccal administration is a parenteral route of administration.

_____ 2. Oral administration is the most frequently used route of administration.

_____ 3. With oral formulations, drugs administered by solid dosage forms generally reach the systemic circulation faster than liquid dosage forms.

_____ 4. The stomach has a pH around 1 to 2.

_____ 5. In an emulsion, if the oleaginous component is present as droplets, the emulsion is called water-in-oil.

_____ 6. Phlebitis can be a complication of intravenous administration.

_____ 7. The needle length for a subcutaneous injection is generally 3/8 to 1 inch.

_____ 8. It takes about 20 seconds for an intravenously administered drug to circulate throughout the body.

_____ 9. Ophthalmic ointment tubes typically hold about 3.5 g of ointment.

_____ 10. Toxic shock syndrome is a disease caused by a bacterial infection.

FILL IN THE KEY TERM

Answers are at the end of the book.

aqueous
atomizer
buccal
buffer system
bulk powders
colloids
diluent
disintegration
dissolution
emulsions
enteric coated
hemorrhoid
hydrates
injectability
intradermal injections
intramuscular injection sites
intravenous sites
IUD
lacrimal canalicula
lacrimal gland
local effect
metered dose inhalers
nasal inhaler
nasal mucosa
necrosis
ophthalmic
parenteral
percutaneous absorption
pH
suspensions
sterile
sublingual
syringeability
syrups
systemic effect
toxic shock syndrome
transcorneal transport
viscosity
water soluble
wheal
Z-tract injection

1. ____________________: When the drug activity is at the site of administration (e.g., eyes, ears, nose, skin).
2. ____________________: When a drug is introduced into the circulatory system by any route of administration and carried to the site of activity.
3. ____________________: Ingredients designed to control the pH of a product.
4. ____________________: Solid formulations to be mixed with water or juice.
5. ____________________: The pouch between the cheeks and teeth in the mouth.
6. ____________________: Absorbs water.
7. ____________________: Coating that will not let the tablet disintegrate until it reaches the higher pHs of the intestine.
8. ____________________: The property of a substance being able to dissolve in water.
9. ____________________: Under the tongue.
10. ____________________: Measure of acidity or alkalinity of a substance.
11. ____________________: Any route that does not involve the alimentary tract.
12. ____________________: Increase in cell death.
13. ____________________: Related to the eye.
14. ____________________: Injections administered into the top layer of the skin using short needles.
15. ____________________: The veins of the antecubital area (in front of the elbow), the back of the hand, and some of the larger veins in the foot.
16. ____________________: Painful/swollen veins in the anal/rectal area.
17. ____________________: Liquid formulations in which the drug does not completely dissolve in the solvent.

18. ____________________ : Tear ducts.
19. ____________________ : Drug transfer into the eye.
20. ____________________ : A mixture of two liquids that do not mix with each other in which one liquid is spread through the other by mixing and use of a stabilizer.
21. ____________________ : Gluteal (buttocks), deltoid (upper arm), and vastus lateralis (thigh) muscles.
22. ____________________ : Ease with which a suspension can be drawn into a syringe.
23. ____________________ : The thickness of a liquid.
24. ____________________ : Water-based.
25. ____________________ : A raised, blister-like area on the skin, as caused by an intradermal injection.
26. ____________________ : The gland that produces tears for the eye.
27. ____________________ : Breaking a part of a tablet into smaller pieces.
28. ____________________ : Suspended formulation with particle size up to a hundred times smaller than a suspension.
29. ____________________ : Injection technique for medications that stain the skin.
30. ____________________ : The cellular lining of the nose.
31. ____________________ : Device used to convert liquid to a spray.
32. ____________________ : A device that contains a drug that is vaporized by inhalation.
33. ____________________ : Ease of flow when a suspension is injected into a patient.
34. ____________________ : Aerosols that use special metering valves to deliver a fixed dose when the aerosol is actuated.
35. ____________________ : The absorption of drugs through the skin, often for a systemic effect.
36. ____________________ : When the smaller pieces of a disintegrated tablet dissolve in solution.
37. ____________________ : Concentrated solutions of sugar in water.
38. ____________________ : Free of microorganisms.
39. ____________________ : A rare and potentially fatal disease that results from a severe bacterial infection of the blood.
40. ____________________ : A solvent that dissolves a freeze-dried powder or dilutes a solution.
41. ____________________ : An intrauterine contraceptive device that is placed in the uterus for a prolonged period of time.

IDENTIFY

Identify the route of administration.

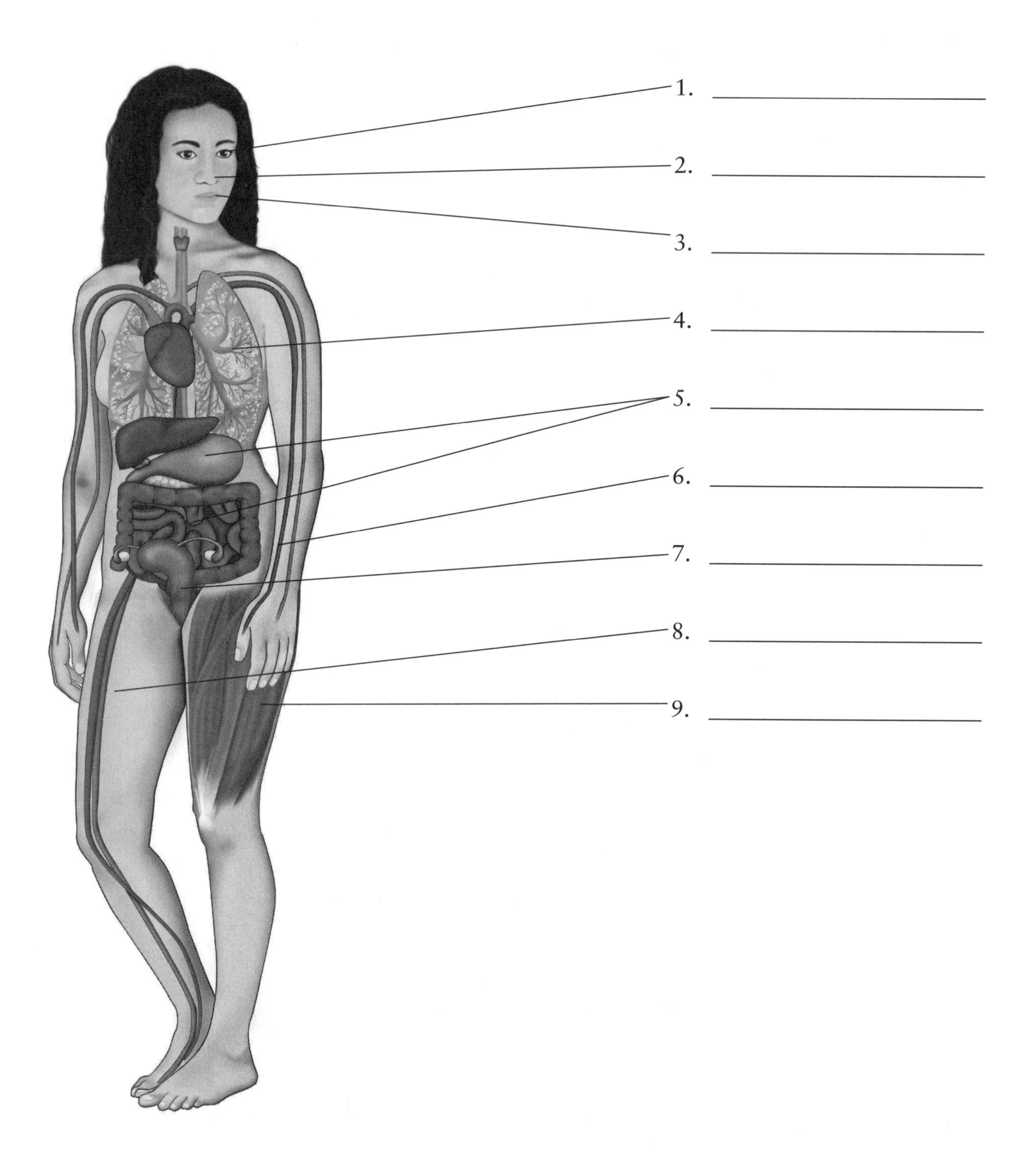

IDENTIFY

Identify the routes of administration.

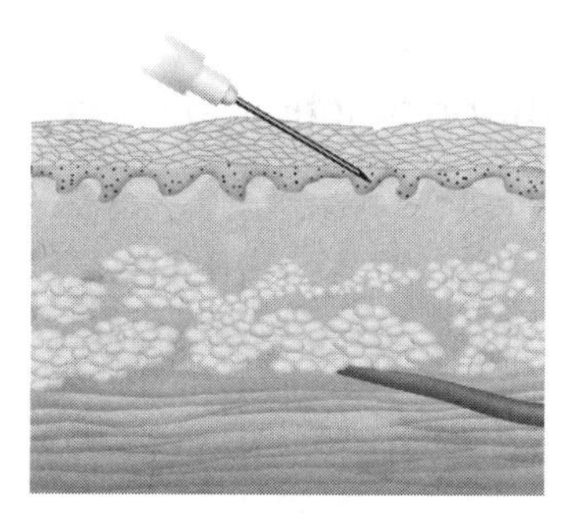

1. ________________

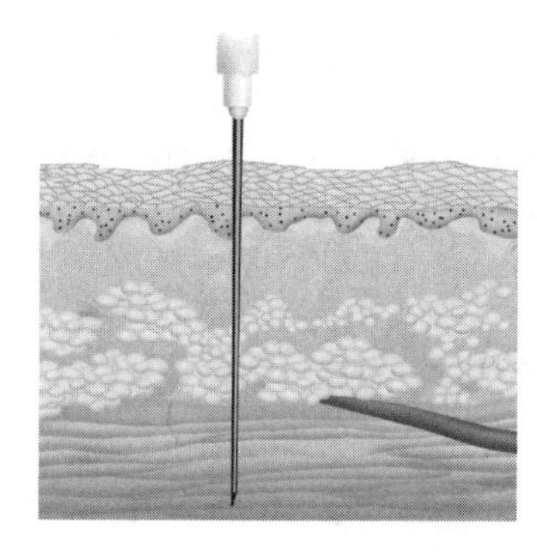

4. ________________

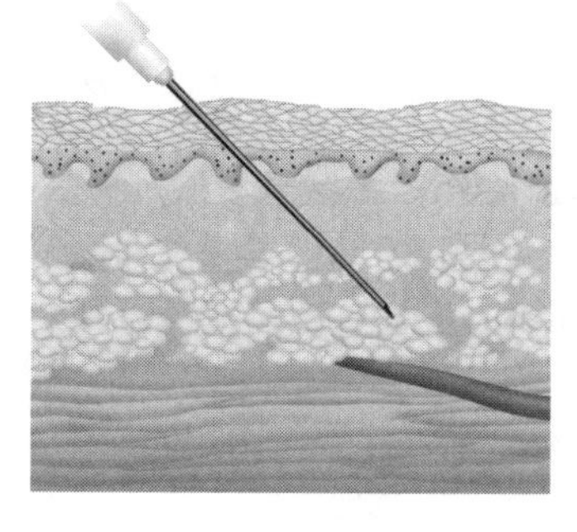

2. ________________

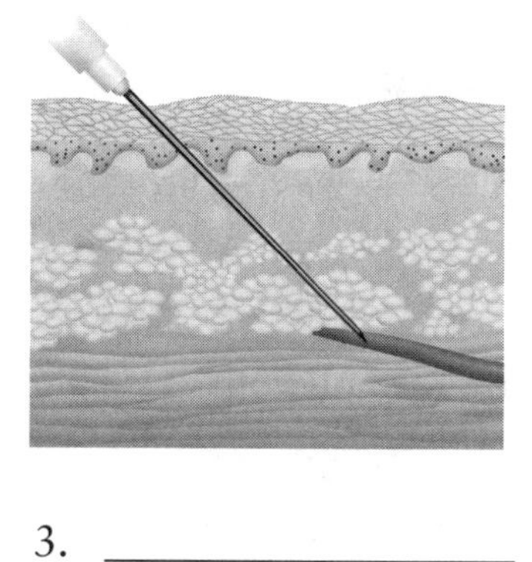

3. ________________

Identify these sites of intramuscular administration on the figure at right:

1. deltoid ____________
2. gluteus maximus ______
3. gluteus medius ______
4. vastus lateralis ______
5. ventrogluteal ______

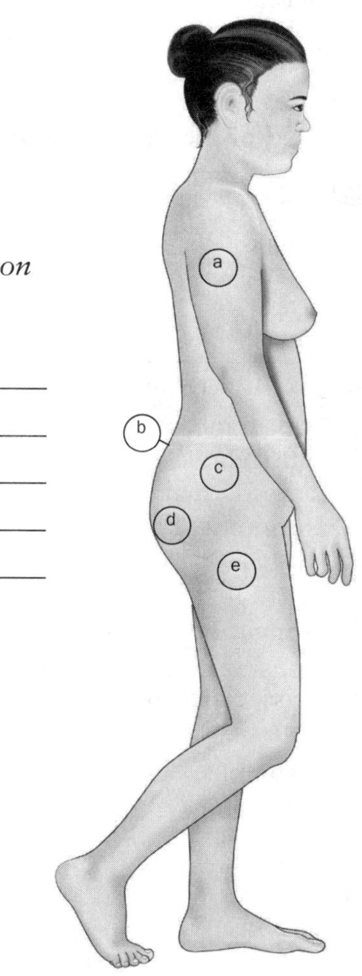

IN THE WORKPLACE ACTIVITIES

1. Print off labels from FDA.gov for the drugs in the prescriptions on pages 70–75 and determine proper storage requirements.
2. Work in pairs. Identify the routes of administration for each drug in the prescriptions on pages 70–75 and discuss reasons why that route of administration would be appropriate.

CHOOSE THE BEST ANSWER

Answers are at the end of the book.

1. ____________________ tablets are placed under the tongue.
   a. Enteric coated
   b. Buccal
   c. Sublingual
   d. TSS

2. The pH of the stomach is around
   a. 1–2.
   b. 4–5.
   c. 5–6.
   d. 6–7.

3. ________________ injections are administered directly into veins.
   a. Subcutaneous
   b. Intravenous
   c. Transdermal
   d. Intramuscular

4. Inflammation of a vein is also known as _______________ and can be a complication associated with intravenous administration.
   a. thrombus
   b. toxic shock
   c. phlebitis
   d. embolus

5. The gradual intravenous injection of a volume of fluid into a patient is called
   a. transdermal.
   b. infiltration.
   c. infusion.
   d. suspension.

6. The most common kind of oral solution:
   a. nonaqueous solutions.
   b. aqueous solutions.
   c. elixirs.
   d. tinctures

7. _________ contain the drug and other ingredients packaged in a gelatin shell.
   a. Tablets
   b. Capsules
   c. Emulsions
   d. Gels

8. ______ contain(s) the active drug in a small powder paper or foil envelope.
   a. Bulk powder
   b. Tablets
   c. Capsules
   d. Modified release formulations

9. Most frequently used as flavoring agents.
   a. Spirits or essences
   b. Tinctures
   c. Elixirs
   d. Suspensions

10. A device that goes between an aerosol's mouthpiece and the patient's mouth is a/an
    a. atomizer.
    b. nebulizer.
    c. spacer.
    d. MDI.

11. ________________ absorption is the absorption of drugs through the skin, often for systemic effect.
    a. Intravenous
    b. Intramuscular
    c. Subcutaneous
    d. Percutaneous

12. A local effect achieved by rectal administration.
    a. Laxative
    b. Analgesic
    c. Antinausea
    d. Anti-infective

13. The most common injection route for insulin is
    a. subcutaneous.
    b. intramuscular.
    c. sublingual.
    d. intradermal.

14. Devices that have special metering valves to administer drugs by inhalation:
    a. spacers
    b. adapters
    c. atomizers
    d. MDI aerosols

STUDY NOTES

Use this area to write important points you'd like to remember.

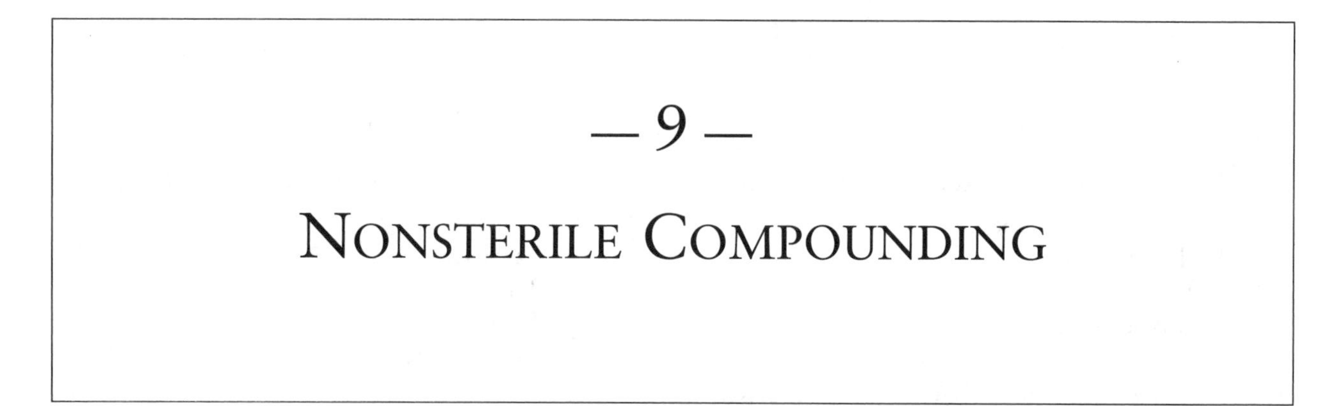

— 9 —

NONSTERILE COMPOUNDING

KEY CONCEPTS

Test your knowledge by covering the information in the right-hand column.

extemporaneous compounding The on-demand formulation of a drug product according to a physician's prescription, formula, or recipe.

United States Pharmacopeia (USP) Establishes standards for pharmacy compounding that are accepted by many states.

quality assurance (QA) Program of activities to assure the compounded formulation meets specifications and satisfies standards.

Formulation Record Formulas and procedures for what should happen when the preparation is compounded.

Compounding Record A record of what actually happened when the preparation was compounded.

Class III torsion balance Can weigh as little as 120 mg of material with a 5% error. Always use the balance on a level surface and in a draft-free area. Always arrest the balance before adding or removing weight from either pan, or storing.

sensitivity The amount of weight that will move the balance pointer one division mark.

weighing papers or boats Should always be placed on the balance pans before any weighing is done. Balances must be readjusted after a new weighing paper or boat has been placed on each pan. Weighing papers taken from the same box can vary in weight by as much as 35 mg.

analytical balances Highly sensitive balances that can weigh quantities smaller than 120 mg with acceptable accuracy.

graduated cylinders Cylindrical graduates are preferred over cone-shaped ones because they are more accurate. When selecting a graduate, always choose the smallest graduate capable of containing the volume to be measured. Avoid measurements of volumes that are below 20% of the capacity of the graduate because the accuracy is unacceptable.

volumetric flasks	For weighing liquid drugs, solvents, or additives. Includes graduates, flasks, pipets, and syringes. Erlenmeyer flasks, beakers, and prescription bottles, regardless of markings, are not volumetric glassware.
syringes	Used to measure small volumes. Measurements made with syringes are more accurate and precise than those made with cylindrical graduates. Measure volumes to the edge of the syringe stopper.
pipets	Devices that help compounders measure a variety of volumes. Pipets are thin glass or plastic tubes for delivering volumes less than 25 mL.
dispensers	Automated filling systems that are often used as fluid delivery devices.
meniscus	The curved surface of a volume of liquid. When reading a volume of a liquid against a graduation mark, hold the graduate so the meniscus is at eye level and read the mark at the bottom of the meniscus.
droppers	Used to deliver small liquid doses, but must first be calibrated.
mortar and pestle	Made of three types of materials: glass, Wedgewood, and porcelain. Wedgewood and porcelain mortars are used to grind crystals and large particles into fine powders. Glass mortars and pestles are preferable for mixing liquids and semisolid dosage forms.
geometric dilution	A technique for mixing two powders of unequal size. The smaller amount of powder is diluted in steps by additions of the larger amount of powder.
trituration	The fine grinding of a powder.
levigation	The trituration of a powdered drug with a solvent in which the drug is insoluble to reduce the particle size of the drug.
aqueous solution	Clear liquids in which the drug is completely dissolved in water.
syrup	A concentrated or nearly saturated solution of sucrose in water. Syrups containing flavoring agents are known as flavoring syrups (e.g., Cherry Syrup, Acacia Syrup, etc.).
nonaqueous solutions	Solutions that contain solvents other than water.
suspensions	A "two-phase" compound consisting of a finely divided solid dispersed in a liquid. Most solid drugs are levigated in a mortar to reduce the particle size as much as possible before adding to the vehicle. Common levigating agents are alcohol or glycerin.
flocculating agents	Electrolytes that carry an electrical charge and enhance particle "dispersibility" in a solution.
thickening agents	Reduce the settling (sedimentation rate) of a suspension.

KEY CONCEPTS

Test your knowledge by covering the information in the right-hand column.

emulsion	An unstable system consisting of at least two immiscible (unmixable) liquids, one that is dispersed as small droplets throughout the other, and a stabilizing agent.
oil-in-water (o/w)	An emulsion of oils, petroleum hydrocarbons, and/or waxes with water, where the aqueous phase is generally in excess of 45% of the total weight of the emulsion.
water-in-oil (w/o)	When water or aqueous solutions are dispersed in an oleaginous (oil based) medium, with the aqueous phase constituting less than 45% of the total weight.
emulsifiers	Emulsifiers provide a protective barrier around the dispersed droplets that stabilize the emulsion. Commonly used emulsifiers include: tragacanth, sodium lauryl sulfate, sodium dioctyl sulfosuccinate, and polymers known as the Spans® and Tweens®.
ointments and creams	Ointments are simple mixtures of a drug(s) in an ointment base. A cream is a semisolid emulsion. Oleaginous (oil based) bases generally release substances slowly and unpredictably. Water miscible or aqueous bases tend to release drugs more rapidly.
suppository bases	There are three classes that are based on their composition and physical properties: oleaginous bases, water soluble or miscible bases, and hydrophilic bases.
polyethylene glycols (PEGs)	Popular water soluble bases that are chemically stable, nonirritating, miscible with water and mucous secretions, and can be formulated by molding or compression in a wide range of hardnesses and melting points.
compression molding	A method of preparing suppositories by mixing the suppository base and the drug ingredients and forcing the mixture into a special compression mold.
fusion molding	A method in which the drug is dispersed or dissolved in a melted suppository base. The fusion method can be used with all types of suppositories and must be used with most of them.
capsules	When filling, the smallest capsule capable of containing the final volume is used since patients often have difficulty swallowing large capsules.

USING A BALANCE

Class III Balance

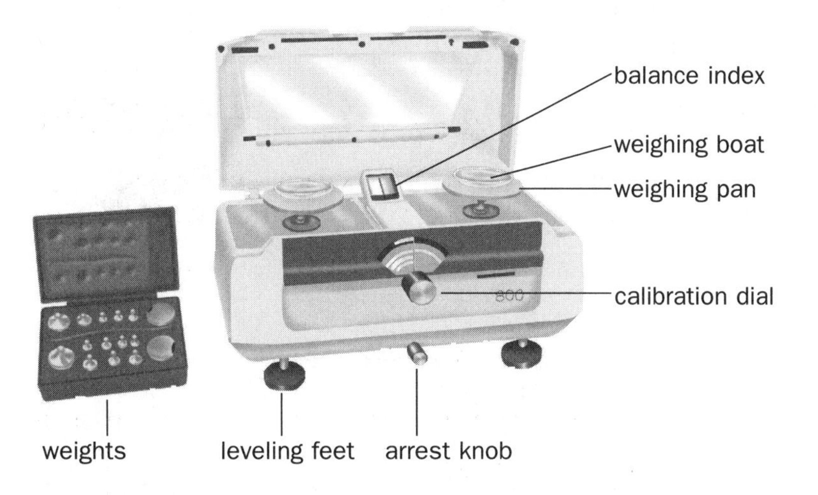

BASIC GUIDELINES FOR USING BALANCES

There are some general rules about using a balance that help to maintain the balance in top condition.

✔ Always cover both pans with weighing papers or use weighing boats. These protect the pans from abrasions, eliminate the need for repeated washing, and reduce loss of drug to porous surfaces.

✔ A clean paper or boat should be used for each new ingredient to prevent cross contamination of components.

✔ The balance must be readjusted after a new weighing paper or boat has been placed on each pan. Weighing papers taken from the same box can vary in weight by as much as 35 mg. If the new zero point is not established, an error of as much as 35 mg can be made. On 200 mg of material, this is more than 15%. Weighing boats also vary in weight.

✔ Always arrest the balance before adding or removing weight from either pan. Although the balance is noted for its durability, repeated jarring of the balance will ultimately damage the working mechanism of the balance and reduce its accuracy.

✔ Always clean the balance, close the lid, and arrest the pans before storing the balance between uses.

✔ Always use the balance on a level surface and in a draft-free area.

MEASURING

MENISCUS

When reading a volume of a liquid against a graduation mark, hold the graduate so the meniscus is at eye level and read the mark at the bottom of the meniscus. Viewing the level from above will create the incorrect impression that there is more volume in the graduate. If the container is very narrow, the meniscus can be quite large.

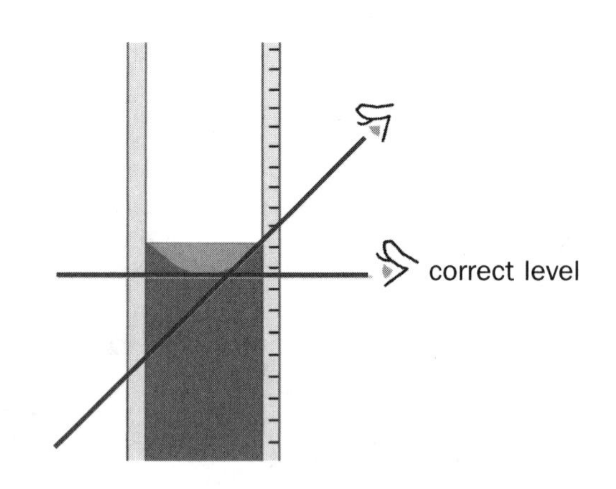

SYRINGE

When reading a volume of a liquid in a syringe, read to the edge of the stopper.

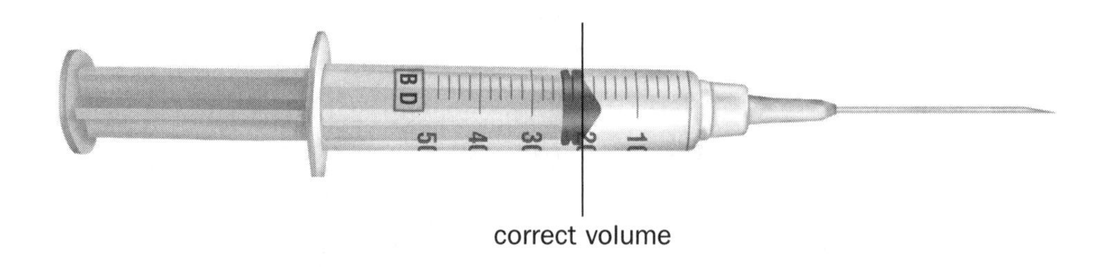

CAPSULE SIZES

The relative sizes and fill capacities of capsules are:

Size	Volume (mL)
000	1.37
00	0.95
0	0.68
1	0.5
2	0.37
3	0.3
4	0.2
5	0.13

the punch method of filling capsules

IN THE WORKPLACE

Sample Compounding Record

Compounding Record

Preparation Name: ______________________

Strength: ______________________

Dosage Form: ______________________

Route of Administration: ______________________

Quantity Prepared: ______________________

Date of Preparation: ______________________

Person Preparing Preparation: ______________________

Person Checking Preparation: ______________________

Ingredient	Manufacturer and Lot Number	Purity Grade	Description	Quantity Required	Actual Quantity Used

Calculations: [Calculations performed at the time of compounding.]

Equipment Operation: [Note any *equipment performance problems or alternate equipment used.*]

Method of Preparation: [Note any deviation from the Formulation Record method of preparation.]

Description of Finished Product:

Quality Control Procedures: [Details of quality assurance test results and data.]

Packaging Container:

Storage Requirements:

Beyond-Use Date Assignment: [What beyond-use date was assigned and reasons for the assignment.]

Auxiliary Label Information:

IN THE WORKPLACE

Sample Formulation Record

Formulation Record

Name: ____________________

Strength: ____________________

Dosage Form: ____________________

Route of Administration: ____________________

Date of Last Review or Revision: ____________________

Person Completing Last Review or Revision: ____________________

Formula:

Ingredient	Quantity	Physical Description	Solubility	Therapeutic Activity

Example Calculations:

Equipment Required:

Method of Preparation:

Description of Finished Product:

Quality Control Procedures:

Packaging Container:

Storage Requirements:

Beyond-Use Date Assignment:

Label Information:

Source of Recipe:

Literature Information:

IN THE WORKPLACE

Sample Standard Operating Procedure Form

Standard Operating Procedure

	Policies and Procedures		
Subject: [What the SOP is concerning]	Effective Date:	Revision Date:	Revision No.:
	Approved by:	Reviewed by:	
	[Additional Items]	[Additional Items]	[Additional Items]

Purpose of Procedure:

[The purpose of the procedure should be described.]

Procedure:

[The procedure should be detailed in a step-by-step fashion so it can be easily followed. It should contain sufficient detail and descriptive information to minimize misinterpretation. It also may contain graphs, charts, figures.]

1.
2.
3.
4.

Documentation:

[Executing the SOP may generate data or information that needs to be documented. This information may be best reported on a form and maintained in a separate notebook or binder. The form should refer to the execution of the SOP by including the name of the procedure, date it was executed, personnel name conducting the procedure, and then the relevant data or information.]

IN THE WORKPLACE ACTIVITIES

1. Determine the correct amount of ingredients for the TPN on page 59 of this book
2. Prepare a compounding record for each of the compounded prescriptions on page 60 of this book.
3. Prepare a formulation record for each of the compounded prescriptions on page 60 of this book.
4. Using a graduated cylinder, demonstrate how to measure volume. Identify the meniscus.
5. Using a syringe, demonstrate how to read volume at the edge of the stopper.
6. Using baking powder, or another powder that is available, and empty gelatin capsules, demonstrate using the punch method of filling capsules. See the diagram on page 100 that illustrates the technique.
7. Using the diagram on page 99 identify the parts of a class III balance.
8. Using the figure on page 100, work in pairs and describe the differences in the different capsule sizes.

TRUE/FALSE

Indicate whether the statement is true or false in the blank. Answers are at the end of the book.

_____ 1. Compounding must always be done upon receipt of a prescription, never in advance.

_____ 2. Protect from freezing means to store above 0°C.

_____ 3. Cylindrical graduates are more accurate than conical ones.

_____ 4. Erlenmeyer flasks are not volumetric glassware.

_____ 5. Disposable syringes are generally used for measuring small volumes.

_____ 6. Transfer pipettes are volumetric devices.

_____ 7. The punch method is a method to prepare suppositories.

_____ 8. When using droppers, all drop sizes are the same regardless of the type of liquid.

_____ 9. Flocculating agents are used to reduce the sedimentation rate in suspensions.

_____ 10. PEGs with molecular weight over 1,000 are solids.

_____ 11. Cocoa butter is a well-known hydrophilic base.

_____ 12. Capsule size 000 can hold more ingredient than capsule size 5.

EXPLAIN WHY

Explain why these statements are true or important. Check your answers in the text. Discuss any questions you may have with your instructor.

1. Why is the stability of a compound important?

2. Why is accuracy in each step of compounding important?

3. Why is the smallest device that will accommodate a volume used to measure it?

4. Why are class A balances not used for very small amounts?

5. Why is geometric dilution used when mixing unequal amounts of powders?

STUDY NOTES

Use this area to write important points you'd like to remember.

FILL IN THE KEY TERM

Answers are at the end of the book.

aliquot
aqueous solutions
beyond-use date
calibrate
compounding record
emulsifier
emulsion
extemporaneous compounding
flocculating agent
formulation record
geometric dilution
hydrophilic emulsifier
immiscible
levigation
lipophilic emulsifier
meniscus
mucilage
nonaqueous solutions
ointments
pipets
primary emulsion
sensitivity
sieves
spatulation
syrup
thickening agent
trituration
USP–NF Chapter <795>
USP–NF Chapter <797>
USP–NF grade
volumetric

1. ____________________ : The on-demand formulation of a drug product according to a physician's prescription, formula, or recipe.
2. ____________________ : Regulations pertaining to nonsterile compounding or formulations.
3. ____________________ : Regulations that pertain to sterile compounding or formulations.
4. ____________________ : To set, mark, or check the graduations of a measuring device.
5. ____________________ : Measures volume.
6. ____________________ : A record of what actually happened when the formulation was compounded.
7. ____________________ : The curved surface of a column of liquid.
8. ____________________ : The fine grinding of a powder.
9. ____________________ : Triturating a powdered drug with a solvent in which it is insoluble to reduce its particle size.
10. ____________________ : A technique for mixing two powders of unequal quantity.
11. ____________________ : Mesh screens.
12. ____________________ : Mixing powders with a spatula.
13. ____________________ : Formulas and procedures of what should happen when a formulation is compounded.
14. ____________________ : A portion of a mixture.
15. ____________________ : A concentrated or nearly saturated solution of sucrose in water.
16. ____________________ : The amount of weight that will move the balance pointer one division mark.

17. ____________________ : Electrolytes used in the preparation of suspensions.

18. ____________________ : An agent used in the preparation of suspensions to increase the viscosity of the liquid.

19. ____________________ : Minimum grade of purity for an ingredient in a compound.

20. ____________________ : Simple mixtures of drug(s) in an ointment base.

21. ____________________ : A stabilizing agent in emulsions.

22. ____________________ : An unstable system consisting of at least two immiscible liquids.

23. ____________________ : Solutions that contain solvents other than water.

24. ____________________ : A date assigned to a compounded prescriptions beyond which the preparation should not be used.

25. ____________________ : A stabilizing agent for water-based dispersion mediums.

26. ____________________ : A stabilizing agent for oil-based dispersion mediums.

27. ____________________ : The initial emulsion formed in a preparation to which ingredients are added to create the final volume.

28. ____________________ : A wet, slimy preparation formed as an initial step in a wet-emulsion preparation method.

29. ____________________ : Clear liquids in which the drug is completely dissolved in water.

30. ____________________ : Cannot be mixed.

31. ____________________ : Thin glass tubes for volumetric measurement.

CHOOSE THE BEST ANSWER

Answers are at the end of the book.

1. Establishes standards of quality, strength, purity, packaging, and labeling for compounded medications:
   a. USP–NF.
   b. FDA.
   c. ASHP.
   d. DEA.

2. The storage temperature definition for a freezer is
   a. -30°C to 0°C.
   b. -20°C to -10°C.
   c. 8°C to 15°C.
   d. 30°C to 40°C.

3. The minimum weighable quantity for a class III balance is
   a. 120 mg.
   b. 500 mL.
   c. 120 mL.
   d. 500 mg.

4. Metric weights used for weighing ingredients using a class III balance should be handled with
   a. water.
   b. fingers.
   c. forceps.
   d. oil.

5. Quantities less than 120 mg may be measured using a/an
   a. arrest
   b. aliquot
   c. calibration
   d. sensitivity

6. ______________________ is the term for triturating a powdered drug with a solvent in which it is insoluble to reduce its particle size.
   a. Suspension
   b. Trituration
   c. Emulsion
   d. Levigation

7. Mixing powders using a spatula is called
   a. extemporaneous compounding.
   b. spatulation.
   c. emulsification.
   d. levigation.

8. The application of sound waves to affect dissolution is
   a. levitation.
   b. sonication.
   c. spatulation.
   d. flocculation.

9. An appropriate flavoring for a metallic tasting drug is
   a. mint.
   b. orange.
   c. cinnamon.
   d. anise.

10. ______________________ are electrolytes used in the preparation of suspensions.
    a. Flocculating agents
    b. Suspending agents
    c. Complex carbohydrates
    d. Simple sugars

11. A two-phase system consisting of a finely divided solid dispersed in a liquid is a/an
    a. suspension.
    b. emulsion.
    c. solution.
    d. trituration.

12. ______________________ are thickening agents used in the preparation of suspensions.
    a. Electrolytes
    b. Preservatives
    c. Flocculating agents
    d. Suspending agents

13. _______ is less sweet than sucrose.
    a. Sorbitol
    b. Saccharin
    c. Aspartame
    d. Stevia

14. The punch method is used to prepare
    a. tablets.
    b. capsules.
    c. suppositories.
    d. emulsions.

15. Polyethylene glycol (PEG) polymers are used to make _______________ suppositories.
    a. cocoa butter
    b. water soluble or miscible
    c. oleaginous
    d. hydrophilic

16. ______________ are mixtures of oleaginous and water miscible bases for making suppositories.
    a. Hydrophobic bases
    b. Hydrophonic
    c. Hydrophilic bases
    d. Hydrotonic bases

STUDY NOTES

Use this area to write important points you'd like to remember.

— 10 —

STERILE COMPOUNDING & ASEPTIC TECHNIQUES

KEY CONCEPTS

Test your knowledge by covering the information in the right-hand column.

parenteral solutions	There are two types of products: large volume parenteral (LVP) solutions and small volume parenteral (SVP) solutions. LVP solutions are typically bags or bottles containing larger volumes of intravenous solutions. SVP solutions are generally contained in ampules or vials.
properties	Solutions for injection or infusion must be sterile, free of visible particulate material, pyrogen-free, stable for their intended use, have a pH around 7.4, and in most (but not all) cases isotonic.
USP <797>	Standards from the United States Pharmacopeia for pharmaceutical compounding of sterile preparations.
Drug Quality & Security Act (DQSA)	Legislation passed by the United States Congress that affects compounding practice.
admixtures	When a drug is added to a parenteral solution, the drug is referred to as the additive, and the final mixture is referred to as the admixture.
total parenteral nutrition solutions	These are complex admixtures composed of dextrose, fat, protein, electrolytes, vitamins, and trace elements. They are hypertonic solutions. Most of the volume of TPN solutions is made up of macronutrients: amino acid solution (a source of protein) and a dextrose solution (a source of carbohydrate calories). Several electrolytes, trace elements, and multiple vitamins (together referred to as micronutrients) may be added to the base solution to meet individual patient requirements. Common electrolyte additives include sodium chloride (or acetate), potassium chloride (or acetate), calcium gluconate, magnesium sulfate, and sodium (or potassium) phosphate. Multiple vitamin preparations containing both water-soluble and fat-soluble vitamins are usually added on a daily basis. A trace element product containing zinc, copper, manganese, selenium, and chromium may be added.

IV fat emulsions Intravenous fat (lipid) emulsion is required as a source of essential fatty acids. It is also used as a concentrated source of calories. Fat provides nine calories per gram, compared to 3.4 calories per gram provided by dextrose. Intravenous fat emulsion may be admixed into the parenteral nutrition solution with amino acids and dextrose, or piggybacked into the administration line.

peritoneal dialysis solutions Used by patients who do not have functioning kidneys to remove toxic substances, excess body waste, and serum electrolytes through osmosis. The solution is administered directly into the peritoneal cavity (the cavity between the abdominal lining and the internal organs) to remove toxic substances, excess body waste, and serum electrolytes through osmosis. These solutions are hypertonic to blood so the water will not move into the circulatory system.

flow rate The rate at which the solution is administered to the patient.

administration devices Administration sets, volume control chambers, IV catheters, and positive pressure pumps are examples of items that are used for administering intravenous preparations.

piggybacks Small volumes of fluid (usually 50–100 mL) infused into the administration set of an LVP solution.

primary engineering controls (PEC) Devices used to provide a classified air environment for preparing sterile products

laminar airflow workstation (LAFW) Establishes and maintains an ultraclean work area for the preparation of IV admixtures.

biological safety cabinets Used in the preparation of hazardous drugs. They protect both personnel and the environment from contamination.

aseptic techniques Maintain the sterility of all sterile items and are used in preparing IV admixtures.

syringes Syringes come in sizes ranging from 1 to 60 mL. As a rule, a syringe size is used that is one size larger than the volume to be measured. The volume of solution in a syringe is measured to the edge of the plunger's stopper while the syringe is held upright and all air has been removed from the syringe.

needle sizes Needle sizes are indicated by length and gauge. The higher the gauge number, the smaller is the lumen (the hollow bore of the needle shaft). Large needles may be needed with highly viscous solutions but are more likely to cause coring.

needle-free devices Devices available for administering pharmaceuticals and for use as transfer devices in compounding that don't have needles

KEY CONCEPTS (CONT'D)

Test your knowledge by covering the information in the right-hand column.

filters Often used to remove contaminating particles from solutions. Depth filters and membrane filters are the two basic groups.

sharps disposal Used syringes and glass items should be disposed of in properly labeled sharps containers.

FILL IN THE KEY TERM

Answers are at the end of the book.

admixture
anhydrous
aseptic techniques
bevel
biological safety cabinets
buffer capacity
clean rooms
compounded sterile preparations
coring
dialysis
diluent
final filter
Flashball
flow rate
gauge
HEPA filter
heparin lock
hypertonic
hypotonic
irrigation solution
ions
isotonic
laminar flow
lyophilized
lumen
membrane filter
osmotic pressure
peritoneal dialysis solution
piggybacks
primary engineering controls
pyrogens
shaft
sharps
TPN solution
valence

1. ______________________ : Techniques that maintain sterile condition.
2. ______________________ : Chemicals produced by microorganisms that can cause pyretic (fever) reactions in patients.
3. ______________________ : Flexible rubber bulb at the end of a needle on an administration set.
4. ______________________ : When a solution has an osmolarity equivalent to another.
5. ______________________ : When a solution has a greater osmolarity than another.
6. ______________________ : When a solution has a lesser osmolarity than another.
7. ______________________ : The rate (in mL/hour or mL/minute) at which the solution is administered to the patient.
8. ______________________ : An injection device that uses heparin to keep blood from clotting in the device.
9. ______________________ : Small volume solutions added to an LVP.
10. ______________________ : Rooms that house laminar flow hoods and biological safety cabinets.
11. ______________________ : The resulting solution when a drug is added to a parenteral solution.
12. ______________________ : Ability of a solution to resist a change in pH.
13. ______________________ : A liquid that dilutes a substance or solution.

14. ____________________ : Without water molecules.

15. ____________________ : An angled surface, as with the tip of a needle.

16. ____________________ : With needles, the higher the number, the thinner the lumen.

17. ____________________ : The hollow center of a needle.

18. ____________________ : When a needle damages the rubber closure of a parenteral container, causing fragments of the closure to fall into the container and contaminate its contents.

19. ____________________ : A filter that attaches to a syringe and filters solution through a membrane as the solution is expelled from the syringe.

20. ____________________ : A compounded sterile dosage fom.

21. ____________________ : A filter that filters solution immediately before it enters a patient's vein.

22. ____________________ : Continuous movement at a stable rate in one direction.

23. ____________________ : A high-efficiency particulate air filter.

24. ____________________ : Are used in the preparation of hazardous drugs and protect both personnel and the environment from contamination.

25. ____________________ : Large splash solutions used during surgical or urologic procedures to bathe and moisten body tissue.

26. ____________________ : The stem of the needle that provides for the overall length of the needle.

27. ____________________ : Needles, jagged glass or metal objects, or any items that might puncture or cut the skin.

28. ____________________ : Molecular particles that carry electrical charges.

29. ____________________ : Characteristic of a solution determined by the number of particles dissolved in it.

30. ____________________ : The number of positive or negative charges on an ion.

31. ____________________ : Freeze dried.

32. ____________________ : A solution placed in and emptied from the peritoneal cavity to remove toxic substances.

33. ____________________ : Intravenous solution with amino acids, dextrose, and additional micronutrients.

34. ____________________ : Devices used to provide a classified air environment.

35. ____________________ : Movement of particles in a solution through permeable membranes.

TRUE/FALSE

Indicate whether the statement is true or false in the blank. Answers are at the end of the book.

_____ 1. Otic dosage forms are not required to be sterile.

_____ 2. 0.9% sodium chloride is an isotonic solution.

_____ 3. Physiological pH is about 7.4.

_____ 4. Coring is more likely to occur with a 27 gauge needle than a 13 gauge needle.

_____ 5. Isopropyl alcohol is a cleaning agent and disinfectant.

_____ 6. A pharmacy must test every compounded preparation for QA.

EXPLAIN WHY

Explain why these statements are true or important. Check your answers in the text. Discuss any questions you may have with your instructor.

1. Why should intravenous solutions generally be isotonic?
2. Why is the position of objects on a laminar flow hood work surface important?
3. Why is visual inspection of parenteral solutions important?

STUDY NOTES

Use this area to write important points you'd like to remember.

Laminar AirFlow Workstations & Biological Safety Cabinets

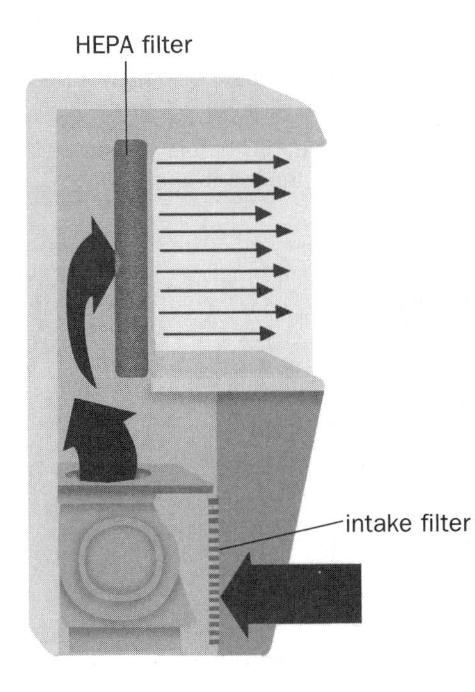

Laminar Airflow Workstations (LAFWs)

Air is drawn into a horizontal or vertical laminar airflow workstation and is passed through a prefilter to remove relatively large contaminants such as dust and lint. The air is then channeled through a high-efficiency particulate air (HEPA) filter that removes particles larger than 0.3 μm (microns). The purified air then flows over the work surface in parallel lines at a uniform velocity (i.e., laminar flow). The constant flow of air from the LAFW prevents room air from entering the work area and removes contaminants introduced in the work area by material or personnel.

top down view

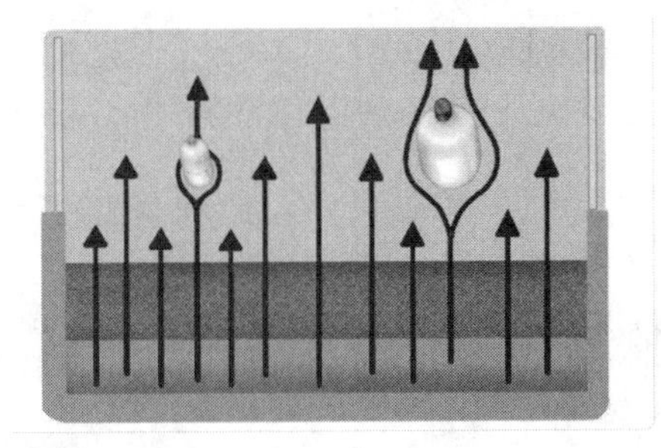

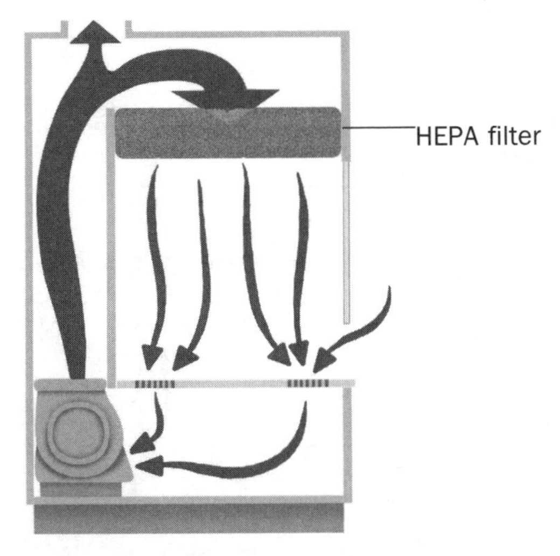

Biological Safety Cabinets

Biological safety cabinets protect both personnel and the environment from contamination. They are used in the preparation of hazardous drugs. A biological safety cabinet functions by passing air through a HEPA filter and directing it down toward the work area. As the air approaches the work surface, it is pulled through vents at the front, back, and sides of the hood. A major portion of the air is recirculated back into the cabinet and a minor portion passes through a secondary HEPA filter and is exhausted into the room.

Rules for Working with LAFWs and Safety Cabinets

- ✔ **Never sneeze, cough, talk directly into an LAFW or cabinet.**
- ✔ **Close doors or windows.** Breezes can disrupt the airflow sufficiently to contaminate the work area.
- ✔ **Perform all work at least 6 inches inside the LAFW or cabinet** to derive the benefits of the laminar airflow. Laminar flow air begins to mix with outside air near the edge of the work area.
- ✔ **Maintain a direct, open path between the filter and the area inside the LAFW/cabinet**
- ✔ **Place nonsterile objects, such as solution containers or your hands, downstream from sterile ones.** Particles blown off these objects can contaminate anything downstream from them.
- ✔ **Do not put large objects at the back of the work area next to the filter.** They will disrupt airflow.

GARBING

When preparing to work in a PEC, the technician should first put on appropriate clothing called "garb." Garbing includes shoe covers, hair bonnets (and beard guards, if applicable), gowns, gloves, and handwashing. These tasks should be completed by covering up from the dirtiest to the cleanest part of one's body. Obviously, dirtiest and cleanest parts are relative terms.

After garbing, the technician carries out any procedures that must be completed in the PEC before compounding begins. Then the technician prepares all the supplies that will be needed and moves them into the PEC

1. Put On Shoe Covers, Hair Bonnet, Mask and Gown

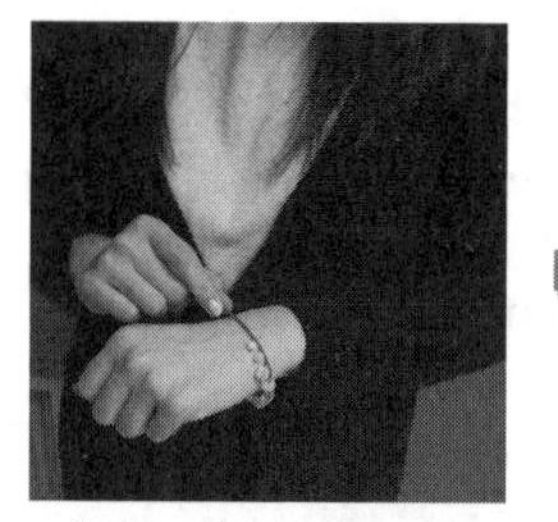

✔ Remove all jewelry.

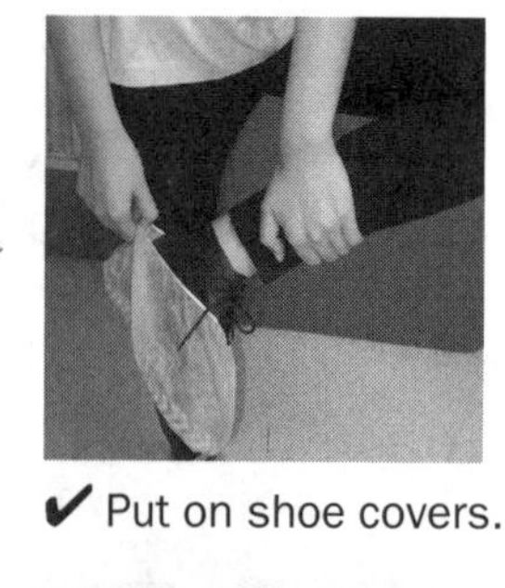

✔ Put on shoe covers.

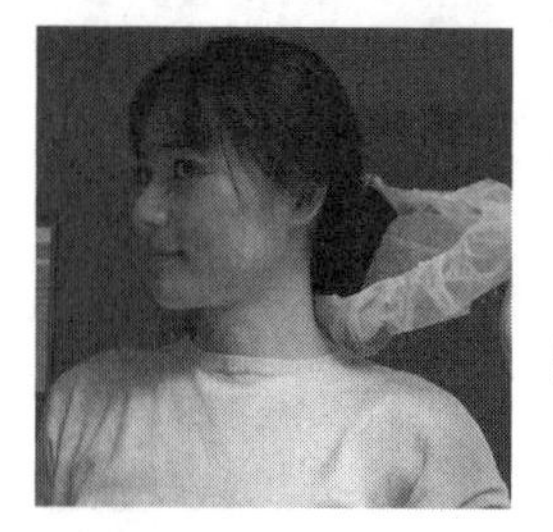

✔ Put on hair bonnet.

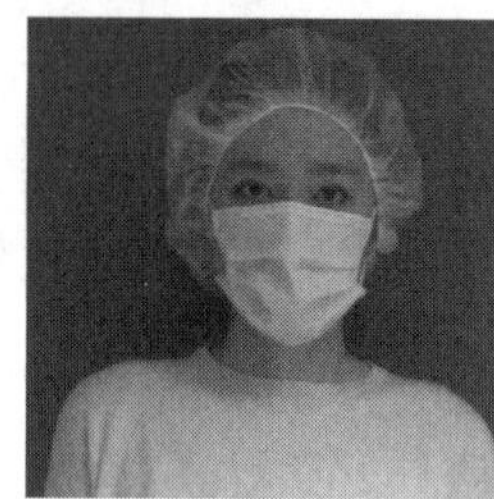

✔ Put on mask.

✔ Put on gown & secure tie strings.

2. Wash Hands

One of the most important ways to minimize contamination is to properly wash hands, following the steps below.

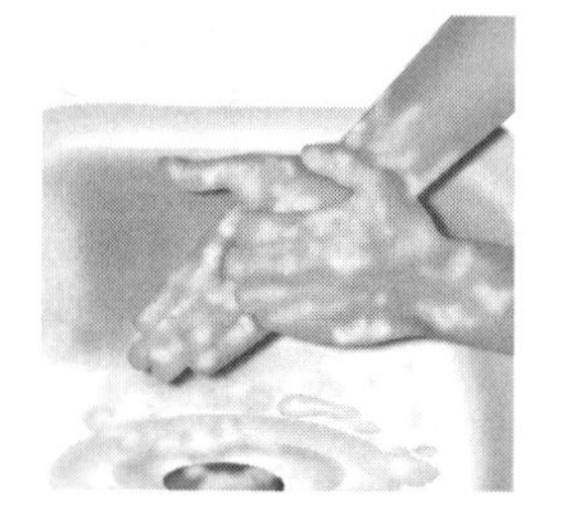

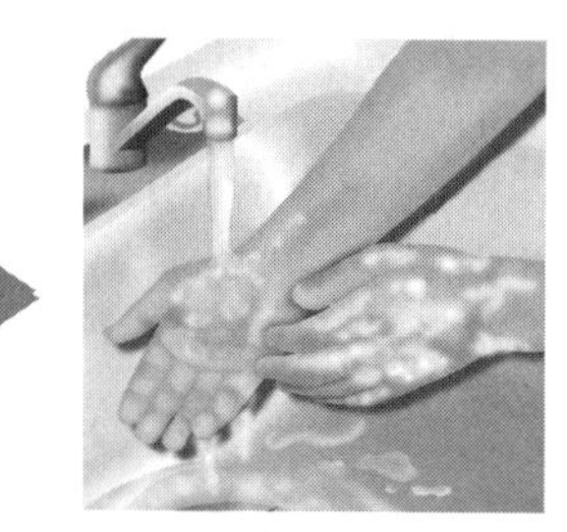

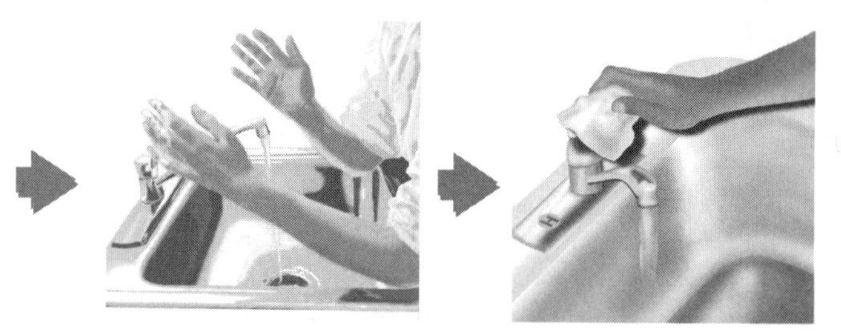

- ✔ Stand far enough away from the sink so clothing does not come in contact with it.
- ✔ Turn on water. Wet hands and forearms thoroughly. Keep hands pointed downward.
- ✔ Scrub hands vigorously with an antibacterial soap.
- ✔ Work soap under fingernails by rubbing them against the palm of the other hand.
- ✔ Interlace the fingers and scrub the spaces between the fingers.
- ✔ Wash wrists and arms up to the elbows with arms pointed upward.
- ✔ Thoroughly rinse the soap from hands and arms with arms pointed upward.
- ✔ Dry hands and forearms thoroughly using a non-shedding paper towel, with arms pointed upward.
- ✔ Use a dry paper towel to turn off the water faucet.
- ✔ After hands are washed, avoid touching clothes, face, hair, or any other potentially contaminated object in the area.

3. Put On Sterile Gloves

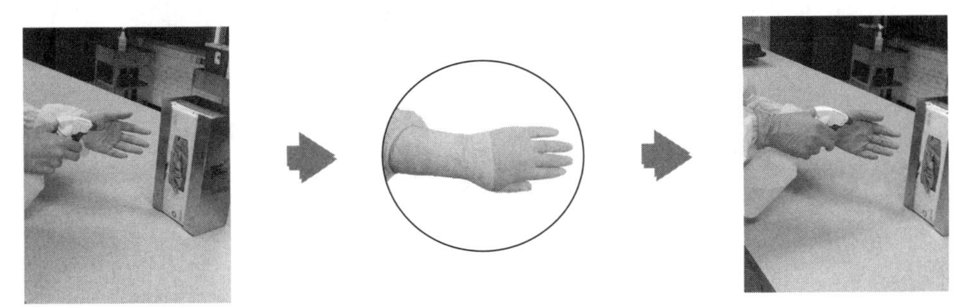

✔ First, spray your hands with sterile 70% isopropyl alcohol. Allow the alcohol to dry.

✔ Then put on sterile gloves, making sure the glove covers the sleeve of the gown.

✔ Finally, spray both gloves with sterile 70% isopropyl alcohol.

In the Workplace Activities

1. Work in pairs and demonstrate the proper technique for adding an SVP into an LVP. Use the skills checklist on page 123.
2. Work in pairs and use the skills checklist on page 120 for preparing a sterile preparation using a medication that is in a glass ampule.
3. Work in pairs and use the skills checklist on page 122 to demonstrate the proper technique to prepare a medication that involves a vial with a powder.
4. Work in pairs and use the skills checklist on page 121 to demonstrate the proper technique for preparing a sterile preparation for one ingredient that is a vial in a solution.
5. Work in pairs and use the skills checklist on page 119 for proper use of a laminar airflow workstation.
6. Simulate appropriate use of protective clothing that should be worn when preparing sterile compounds.
7. Use the diagram of the laminar airflow workstation on page 115 to show the direction of airflow for a horizontal workstation. Next, show the direction of airflow for a vertical workstation. Also identify where the air intake filter is in the diagram.
8. Using the diagram on page 115, identify the location of the HEPA filter in a biological safety cabinet.
9. Work in pairs, and use the skills checklist on page 118 for aseptic technique: handwashing.

IN THE WORKPLACE

Use these tools to practice and check some of your workplace skills.

Pharmacy Technician Skills Checklist
ASEPTIC TECHNIQUE: HAND WASHING

Name: ____________________

Skill or Procedure	Self-Assessment		Supervisor Review		
	Needs to Improve	Meets or Exceeds	Needs to Improve	Meets or Exceeds	Plan of Action
1. Removes all jewelry and scrubs hands and arms to the elbows with suitable antibacterial agent.					
2. Stands far enough away from sink so clothing does not come in contact with sink.					
3. Turns on water, wets hands and forearms thoroughly, keeps hands pointed downward.					
4. Scrubs hands vigorously with antibacterial soap.					
5. Works soap under fingernails by rubbing them against the palm of the other hand.					
6. Interlaces fingers and scrubs the spaces between the fingers.					
7. Washes wrists and arms up to the elbows.					
8. Thoroughly rinses the soap from hands and arms.					
9. Dries hands and forearms thoroughly using a nonshedding paper towel.					
10. Uses a dry paper towel to turn off water faucet.					

Pharmacy Technician Skills Checklist
LAMINAR AIRFLOW WORKSTATION (LAFW)

Name:____

Skill or Procedure	Self-Assessment		Supervisor Review		
	Needs to Improve	Meets or Exceeds	Needs to Improve	Meets or Exceeds	Plan of Action
1. Turns on and lets run for at least 30 minutes prior to use.					
2. Does not allow jewelry, long sleeves, or other non-sterile materials within the hood.					
3. Uses clean gauze/sponge to clean hood with 70% isopropyl alcohol.					
4. Uses long, side-to-side motions and works from top to bottom to clean back surface of workstaions.					
5. Uses back-to-front motions, working from the top to the bottom of each side to clean the sides of the workstation.					
6. Uses back-to-front motions to clean the surface of the workstation.					
7. Takes care so that cleaned surfaces do not become contaminated during cleaning.					
8. Takes care when placing items in workstation so that airflow is not blocked.					
9. Takes care when preparing admixtures, that airflow is not blocked by hands or other objects.					
10. Takes care so that hands remain under the workstation during admixture preparation, and does not leave the workstation during admixture preparation.					
11. Does not utilize outer 6 inches of workstation opening or work too closely to sides and back of workstation during drug preparation and manipulations.					
12. Does not contaminate workstation by coughing, sneezing, chewing gum, or excessive talking.					

IN THE WORKPLACE

Use these tools to practice and check some of your workplace skills.

Pharmacy Technician Skills Checklist
GLASS AMPULES

Name:____

	Self-Assessment		Supervisor Review		
Skill or Procedure	Needs to Improve	Meets or Exceeds	Needs to Improve	Meets or Exceeds	Plan of Action
1. If ampule is not prescored, uses a fine file to lightly score the neck of the ampule at its narrowest point.					
2. Holds ampule upright and taps the top.					
3. Swabs neck of ampule with an alcohol swab.					
4. Wraps gauze pad around neck of ampule and quickly snaps ampule moving hands outward and away.					
5. Inspects opened ampule for glass particles.					
6. Tilts ampule (about 20° angle).					
7. Inserts needle into ampule, so needle point does not touch opening of ampule.					
8. Positions needle into solution placing beveled edge against side of ampule.					
9. Withdraws correct amount of the drug while keeping the needle submerged.					
10. Withdraws needle from ampule and removes air bubbles from syringe					
11. Transfers solution to final container using filter needle or membrane filter.					

Pharmacy Technician Skills Checklist
VIAL WITH SOLUTION

Name:______________________________

	Self-Assessment		Supervisor Review		
Skill or Procedure	Needs to Improve	Meets or Exceeds	Needs to Improve	Meets or Exceeds	Plan of Action
1. Takes care in removing vial cover.					
2. Uses care in cleaning top of vial with alcohol wipe.					
3. Draws into syringe a volume of air equal to the volume of drug to be withdrawn.					
4. Penetrates the vial without coring and injects air.					
5. Turns the vial upside down and withdraws correct amount of drug into syringe.					
6. Withdraws needle from vial and with needle end up, taps syringe to allow air bubbles to come to the top of the syringe. Presses plunger to remove air and excess solution.					
7. Transfers solution into the IV bag or bottle, minimizing coring.					

IN THE WORKPLACE

Use these tools to practice and check some of your workplace skills.

Pharmacy Technician Skills Checklist
VIAL WITH POWDER

Name:____________________

Skill or Procedure	Self-Assessment		Supervisor Review		
	Needs to Improve	Meets or Exceeds	Needs to Improve	Meets or Exceeds	Plan of Action
1. Takes care in removing vial cover.					
2. Uses care in cleaning top of vial with alcohol wipe.					
3. Draws into syringe a volume of air equal to the volume of diluent to be withdrawn.					
4. Penetrates the diluent vial without coring and injects air.					
5. Turns the diluent vial upside down and withdraws correct amount of diluent into syringe.					
6. Injects diluent into medication vial, and then withdraws a slight amount of air.					
7. Shakes vial until drug dissolves (unless shaking is not recommended)					
8. Reinserts needle and removes proper volume of drug solution (without injecting air).					
9. Removes all bubbles from syringe and transfers reconstituted solution to final container.					

Pharmacy Technician Skills Checklist
ADDING A DRUG (SVP) TO AN LVP

Name: ____________________

	Self-Assessment		Supervisor Review		
Skill or Procedure	Needs to Improve	Meets or Exceeds	Needs to Improve	Meets or Exceeds	Plan of Action
1. Removes protective covering from LVP package.					
2. Assembles the needle and syringe.					
3. If drug is in powder form, reconstitutes drug with recommended diluent.					
4. Swabs the SVP with an alcohol swab and draws the necessary volume of drug solution.					
5. Swabs the medication port of the LVP with an alcohol swab.					
6. Inserts needle into the medication port and through the inner diaphragm (medication port is fully extended).					
7. Injects the SVP solution.					
8. Removes the needle.					
9. Shakes and inspects the admixture.					

CHOOSE THE BEST ANSWER

Answers are at the end of the book.

1. Pyrogens are chemicals that are produced by
   a. coring.
   b. microorganisms.
   c. precipitation.
   d. heat.

2. A(an) ______________ solution has lower osmolarity than blood.
   a. hypertonic
   b. isotonic
   c. hypotonic
   d. pyrogenic

3. Minibags usually contain ___________ of fluid and are infused over a period of 30–60 minutes.
   a. 25–100 mL
   b. 250–500 mL
   c. 5,000–1,000 mL
   d. 1,000 mL–2,000 mL

4. When a drug is added to a parenteral solution, the drug is referred to as the _____________ and the final mixture is referred to as the ________________.
   a. admixture, additive
   b. suspension, solution
   c. additive, admixture
   d. solution, suspension

5. The type of compounding isolator that provides protection for the worker.
   a. Compounding aseptic isolator
   b. Compounding aseptic containment isolator
   c. Baker SteriSHIELD
   d. Baker SeriGUARD

6. _____________ is the part of the needle that attaches to the syringe.
   a. The lumen
   b. The bevel
   c. The coring
   d. The hub

7. _______________ are filters that can be placed inside of needles.
   a. Membrane filters
   b. Filter needles
   c. HEPA filters
   d. Intake filters

8. In horizontal laminar flow hoods, air blows
   a. down toward the work area.
   b. away from the operator.
   c. toward the operator.
   d. up toward the HEPA filter.

9. Biological safety cabinets have
   a. vertical air low down toward the work.
   b. horizontal airflow away from the operator.
   c. vertical airflow up toward the HEPA filter.
   d. horizontal airflow toward the operator.

10. When positioning supplies for use in a laminar flow hood
    a. larger supplies should be placed closer to the HEPA filter.
    b. smaller supplies should be placed closer to the HEPA filter.
    c. space the supplies close together to minimize laminar flow.
    d. the spacing of supplies has no effect on laminar flow.

11. Sharps containers should be disposed of when _____ full.
    a. 1/3
    b. 1/2
    c. 2/3
    d. 3/4

12. If an ampule has not been prescored, the technician should
    a. score the neck of the ampule with a fine file.
    b. use sterile pliers to open the ampule.
    c. use a file to file all the way through the neck.
    d. just snap the ampule since manufacturers always prescore when necessary.

13. If 500 mg of a drug are added to a 1,000 mL IV bag, what is the percent strength of the resulting solution?
    a. 0.0005%
    b. 0.005%
    c. 0.05%
    d. 0.5%

14. What is the molecular weight of $CaCl_2$ if the atomic weight of Ca is 40.08 and the atomic weight of Cl is 35.43?
    a. 70.86
    b. 110.94
    c. 75.51
    d. 106.29

15. What is the valence of KCl?
    a. 0
    b. 2
    c. 1
    d. 3

16. What is the weight of one osmole of KCl? (molecular weight = 74.6)
    a. 74.6 g
    b. 74.6 mg
    c. 37.3 g
    d. 37.3 mg

STUDY NOTES

Use this area to write important points you'd like to remember.

— 11 —

BASIC BIOPHARMACEUTICS

KEY CONCEPTS

Test your knowledge by covering the information in the right-hand column.

objective of drug therapy — To deliver the right drug, in the right concentration, to the right site of action at the right time to produce the desired effect.

receptors — When a drug produces an effect, it is interacting on a molecular level with cell material that is called a receptor. Receptor activation is responsible for most of the pharmacological responses in the body.

site of action — Only those drugs able to interact with the receptors in a particular site of action can produce effects in that site. This is why specific cells only respond to certain drugs.

agonists — Drugs that activate receptors and produce a response that may either accelerate or slow normal cell processes.

antagonists — Drugs that bind to receptors but do not activate them. They prevent other drugs or substances from interacting with receptors.

dose-response curve — Specific doses of a drug is given to various subjects and the effect or response is measured in terms of dose and effect.

blood concentration — The primary way to monitor a drug's concentration in the body and its related effect is to determine its blood concentration.

minimum effective concentration (MEC) — When there is enough drug at the site of action to produce a response.

minimum toxic concentration (MTC) — An upper blood concentration limit beyond which there are undesired or toxic effects.

therapeutic window	The range between the minimum effective concentration and the minimum toxic concentration is called the therapeutic window. When concentrations are in this range, most patients receive the maximum benefit from their drug therapy with a minimum of risk.
ADME	Blood concentrations are the result of four simultaneously acting processes: absorption, distribution, metabolism, and excretion.
disposition	Another term for ADME.
elimination	Metabolism and excretion combined.
half-life	The amount of time it takes for the blood concentration of a drug to decline to one-half of an initial value.
passive diffusion	Besides the four ADME processes, a critical factor of drug concentration and effect is how drugs move through biological membranes. Most drugs penetrate biological membranes by passive diffusion.
hydrophobic drugs	Lipid (fat) soluble drugs that penetrate the lipoidal (fat-like) cell membrane better than hydrophilic drugs.
hydrophilic drugs	Drugs that are attracted to water.
lipoidal	Fat-like or lipid loving.
absorption	The transfer of drug into the blood from an administered drug product is called absorption.
gastric emptying	Most drugs are given orally and absorbed into the blood from the small intestine. One of the primary factors affecting oral drug absorption is the gastric emptying time.
distribution	The movement of a drug within the body once the drug has reached the blood.
selective action	Drug action that is selective to certain tissues or organs, due both to the specific nature of receptor action as well as to various factors that can affect distribution.
protein binding	Many drugs bind to proteins in blood plasma to form a complex that is too large to penetrate cell openings. So the drug remains inactive.

metabolism — The body's process of transforming drugs. The primary site of drug metabolism in the body is the liver. Enzymes produced by the liver interact with drugs and transform them into metabolites.

enzyme — A complex protein that causes chemical reactions in other substances.

metabolite — The transformed drug.

enzyme induction — The increase in enzyme activity that results in greater metabolism of drugs.

enzyme inhibition — The decrease in enzyme activity that results in reduced metabolism of drugs.

first-pass metabolism — When a drug is substantially degraded or destroyed by the liver's enzymes before it reaches the circulatory system, an important factor with orally administered drugs.

enterohepatic cycling — The transfer of drugs and their metabolites from the liver to the bile in the gallbladder and then into the intestine.

excretion — The process of excreting drugs and metabolites, primarily performed by the kidney through the urine.

glomerular filtration — The blood-filtering process of the kidneys. As plasma water moves through the nephron, waste substances (including drugs and metabolites) are secreted into the fluid, with urine as the end result.

bioavailability — The amount of a drug that is delivered to the site of action and the rate at which it is available is called the bioavailability of the drug.

bioequivalency — The comparison of bioavailability between two dosage forms.

pharmaceutical equivalents — Pharmaceutical equivalents are drug products that contain identical amounts of the same active ingredients in the same dosage form, but may contain different inactive ingredients.

pharmaceutical alternatives — Pharmaceutical alternatives are drug products that contain the identical active ingredients, but not necessarily in the same amount or dosage form.

therapeutic equivalent — Pharmaceutical equivalents that produce the same effects in patients.

DOSE RESPONSE CURVE

When a series of specific doses is given to a number of people, the results show that some people respond to low doses but others require larger doses for a response to be produced. Some differences are due to the product itself, but most are due to human variability: different people have different characteristics that affect how a drug product behaves in them. A dose-response curve shows that as doses increase, responses increase up to a point where increased dosage no longer results in increased response.

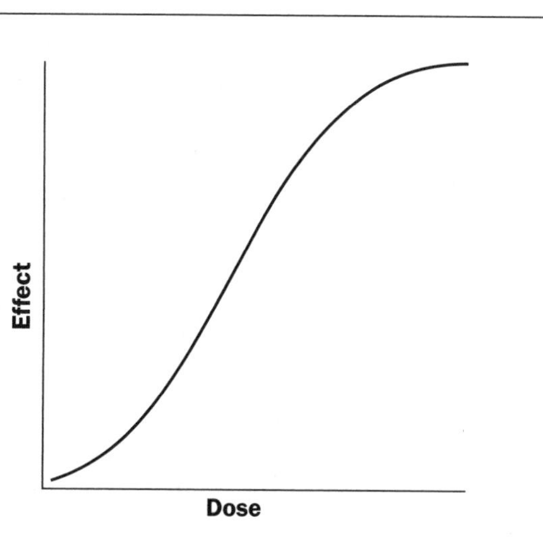

BLOOD CONCENTRATION—TIME PROFILES

Blood concentration begins at zero at the time the drug is administered (before it has been absorbed into the blood). With time, the drug leaves the formulation and enters the blood, causing concentrations to rise. Minimum effective concentration (MEC) is when there is enough drug at the site of action to produce a response. The time this occurs is called the onset of action. With most drugs, when blood concentrations increase, so does the intensity of the effect, since blood concentrations reflect the site of action concentrations that produce the response.

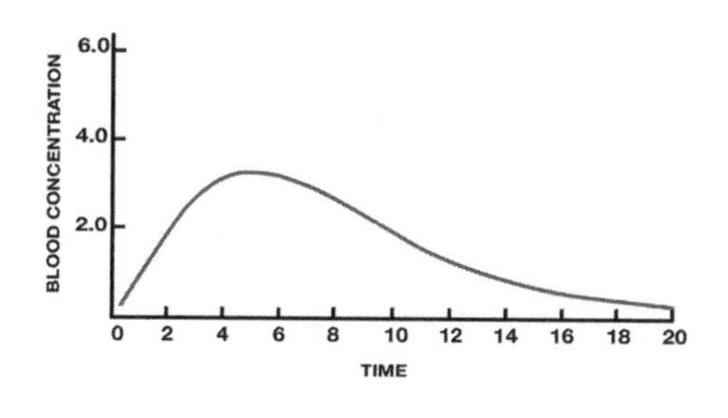

Some drugs have an upper blood concentration limit beyond which there are undesired or toxic effects. This limit is called the minimum toxic concentration (MTC). The range between the minimum effective concentration and the minimum toxic concentration is called the therapeutic window. When concentrations are in this range, most patients receive the maximum benefit from their drug therapy with a minimum of risk.

The last part of the curve shows the blood concentrations declining as absorption is complete. The time between the onset of action and the time when the minimum effective concentration is reached by the declining blood concentrations is called the duration of action. The duration of action is the time the drug should produce the desired effect.

ORAL ABSORPTION

Most drugs are given orally and absorbed into the blood from the small intestine. The small intestine's large surface area makes absorption easier. However, there are many conditions in the stomach that can affect absorption positively or negatively before the drug even reaches the small intestine. One of the primary factors affecting oral drug absorption is the gastric emptying time. This is the time a drug will stay in the stomach before it is emptied into the small intestine. Since stomach acid can degrade many drugs and since most absorption occurs in the intestine, gastric emptying time can significantly affect a drug's action. If a drug remains in the stomach too long, it can be degraded or destroyed, and its effect decreased. Gastric emptying time can be affected by a various conditions, including the amount and type of food in the stomach, the presence of other drugs, the person's body position, and their emotional condition. Some factors increase the gastric emptying time, but most slow it. The pH of the gastrointestinal organs is illustrated at right.

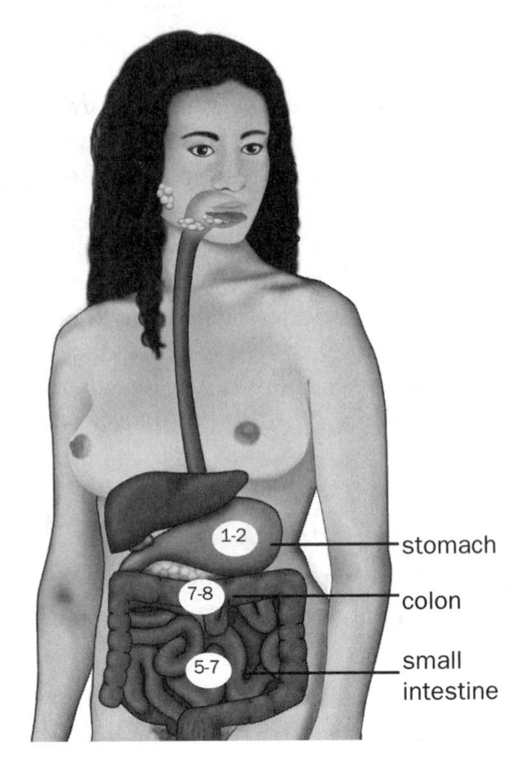

PASSIVE DIFFUSION

Before an effective concentration of a drug can reach its site of action, it must overcome many barriers, most of which are biological membranes, complex structures composed of lipids (fats) and proteins. Most drugs penetrate biological membranes by passive diffusion. This occurs when drugs in the body's fluids move from an area of higher concentration to an area of lower concentration, until the concentrations in each area are in a state of equilibrium. Passive diffusion causes most orally administered drugs to move from the intestine to the blood and from the blood to the site of action.

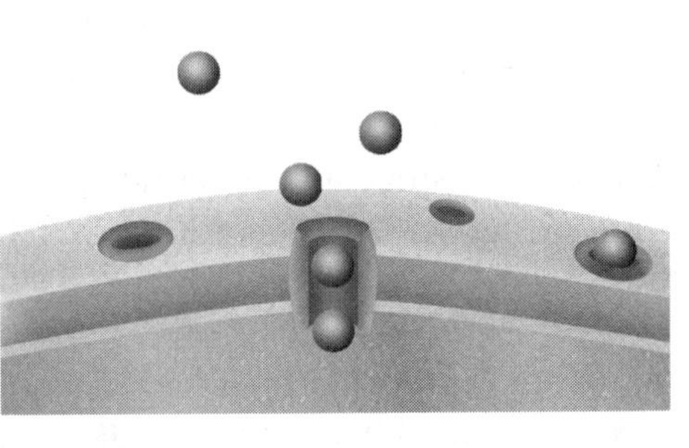

PROTEIN BINDING

Many drugs bind to proteins in blood plasma to form a complex that is too large to penetrate cell openings. So the drug remains inactive. Protein binding can be considered a type of drug storage within the body. Some drugs bind extensively to proteins in fat and muscle, and are gradually released as the blood concentration of the drug falls. These drugs remain in the body a long time, and therefore have a long duration of action.

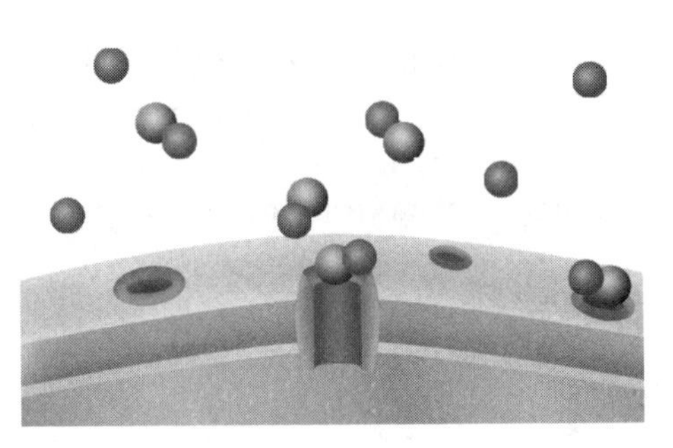

PHARMACOKINETICS (HALF-LIFE)

Besides providing correlations between a pharmacological effect and time, blood concentration–time profiles can be used to describe rate processes. The rate of transfer of an amount of drug from one location to another (e.g., drug transferred to the urine by urinary excretion) or from one chemical form to another (e.g., metabolism of active drug to inactive drug) are common examples of rate processes. Pharmacokinetics is the study of the ADME processes of the body that affect an administered drug.

The data from a blood concentration–time profile is used to determine the bioavailability and bioequivalency of drug products. The same data can be used to obtain three additional pharmacokinetic parameters: half-life of elimination, area-under-curve (AUC), and total body clearance.

HALF-LIFE OF ELIMINATION: This half-life is the amount of time it takes for a blood concentration to decline to one-half an initial value. For example, if at 6 hours, a blood concentration was 30 mcg/mL, and at 15 hours, the blood concentration was 15 mcg/mL, then the half-life would be equal to 9 hours (15 – 6 hours). For most cases, five times the half-life of elimination will estimate how long the elimination processes will be important in the disposition of the administered drug. In the example, 5 x 9 hours, or 45 hours, is low long it will take for the processes of elimination to essentially remove the drug from the body.

AREA-UNDER-CURVE: The AUC is a mathematical number expressing how much physical area is contained under the blood concentration–time profile. The units of AUC are concentration x time (mcg/ml x hours). The AUCs of formulations are used to help determine the bioequivalency of different dosage forms. It is also used to calculate another important pharmacokinetic parameter, total body clearance.

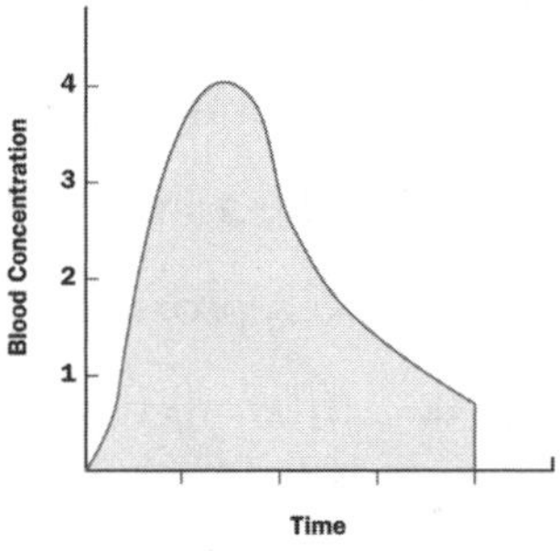

TOTAL BODY CLEARANCE: Total body clearance (Cl) reflects the combined effect of all processes eliminating the drug from the body: urinary excretion, metabolism, expiration through the lungs, and fecal excretion. The units of Cl are mL/min or mL/hour: clearance tells you the number of ml cleared of the drug per time unit. For example, if a drug's clearance is 120 mL/min, then 120 mL of volume is cleared of the drug every minute. That can be said from a conceptual point of view. But in the body, you don't just clear a portion of the volume. Distribution is an ongoing process, as is elimination, so any volume that is cleared will be "refilled" as soon as it is emptied. The overall effect is that blood concentrations continue to fall as the drug is cleared from the body.

IN THE WORKPLACE ACTIVITIES

1. Using the diagram on page 129 show the point in the curve where the maximum blood concentration is achieved.
2. Work in pairs. Using the blood concentration–time profile diagram on page 129 estimate the blood concentration it two minutes, four minutes, 10 minutes, and 20 minutes.
3. Work in pairs, and look at the dose response curve on page 129. Identify the part in the curve where the maximum affect is achieved.
4. Work in pairs. Using the diagram on page 130, identify the part of the gastrointestinal tract where most absorption takes place. Identify the part of the gastrointestinal tract where the pH is 1 to 2. Identify the part of the gastrointestinal tract where the pH is 7 to 8. Identify the part of the gastrointestinal tract with the pH is 5 to 7.
5. Working in pairs, draw a diagram to illustrate passive diffusion.
6. Working in pairs, draw a diagram to illustrate protein binding.

TRUE/FALSE

Indicate whether the statement is true or false in the blank. Answers are at the end of the book.

_____ 1. Receptors are located on the surfaces of cell membranes and inside cells.

_____ 2. Like a lock and key, only certain drugs are able to interact with certain receptors.

_____ 3. Receptors can be changed by drug use.

_____ 4. After all receptors are occupied by a drug, its effect can still be increased by increasing the dose.

_____ 5. When acids dissociate they become unionized.

_____ 6. Most patients receive the maximum benefit from drug therapy when the amount of the drug in the blood is between the minimum effective and minimum toxic concentration.

_____ 7. Five times the half-life is used to estimate how long it takes for elimination from the body.

_____ 8. Concentrations decrease during absorption, before absorption reaches complexation.

_____ 9. Most orally administered drugs are absorbed from the stomach.

_____ 10. Protein binding can result in the gradual release of a drug into the bloodstream.

_____ 11. Erythromycin capsules and erythromycin tablets are pharmaceutical equivalents.

EXPLAIN WHY

Explain why these statements are true or important. Check your answers in the text. Discuss any questions you may have with your instructor.

1. Why can a drug be developed to have a specific therapeutic effect?

2. Why can the same drug have different effects in different people?

3. Why is gastric emptying time important?

4. Why does the chronic administration of some drugs require increases in dosage to achieve the same effect or decreases to avoid toxicity?

5. Why is a pharmaceutical equivalent not necessarily therapeutically equivalent?

STUDY NOTES

Use this area to write important points you'd like to remember.

FILL IN THE KEY TERM

Answers are at the end of the book.

absorption
active transport
agonist
antagonist
bioavailability
bioequivalency
biopharmaceutics
complexation
disposition
duration of action
elimination
enterohepatic cycling
enzyme
enzyme induction
enzyme inhibition
first pass metabolism
gastric emptying time
hydrophilic
hydrophobic
ionized
lipoidal
metabolite
nephron
passive diffusion
pharmaceutical alternative
pharmaceutical equivalent
protein binding
receptor
therapeutic equivalent
therapeutic window
unionized

1. ____________________ : The study of the factors associated with drug products and physiological processes, and the resulting systemic concentrations of the drugs.
2. ____________________ : The process of metabolism and excretion.
3. ____________________ : The cellular material at the site of action that interacts with the drug.
4. ____________________ : The movement of the drug from the dosage form to the blood.
5. ____________________ : Drugs that activate receptors to accelerate or slow normal cell function.
6. ____________________ : Drugs that bind with receptors but do not activate them. They block receptor action by preventing other drugs or substances from activating them.
7. ____________________ : When different molecules associate or attach to each other.
8. ____________________ : The time the drug concentration is above the MEC.
9. ____________________ : A drug's blood concentration range between its minimum effective concentration and minimum toxic concentration.
10. ____________________ : A term sometimes used to refer to all of the ADME processes together.
11. ____________________ : The movement of drugs from an area of higher concentration to lower concentration.
12. ____________________ : The movement of drug molecules across membranes by active means, rather than passive diffusion.
13. ____________________ : Water repelling; cannot associate with water.

14. ____________________ : Capable of associating with or absorbing water.

15. ____________________ : Fat-like substance.

16. ____________________ : The time a drug will stay in the stomach before it is emptied into the small intestine.

17. ____________________ : When acids dissociate.

18. ____________________ : The attachment of a drug molecule to a plasma or tissue protein, effectively making the drug inactive, but also keeping it within the body.

19. ____________________ : The substance resulting from the body's transformation of an administered drug.

20. ____________________ : A complex protein that causes chemical reactions in other substances.

21. ____________________ : The increase in enzyme activity that results in greater metabolism of drugs.

22. ____________________ : The decrease in enzyme activity that results in reduced metabolism of drugs.

23. ____________________ : The substantial degradation of a drug caused by enzyme metabolism in the liver before the drug reaches the systemic circulation.

24. ____________________ : The transfer of drugs and their metabolites from the liver to the bile in the gallbladder and then into the intestine.

25. ____________________ : The functional unit of the kidneys.

26. ____________________ : Drugs that are more lipid soluble.

27. ____________________ : The relative amount of an administered dose that reaches the general circulation and the rate at which this occurs.

28. ____________________ : The comparison of bioavailability between two dosage forms.

29. ____________________ : Drug products that contain identical amounts of the same active ingredients in the same dosage form.

30. ____________________ : Drug products that contain the same active ingredients, but not necessarily in the same amount or dosage form.

31. ____________________ : Pharmaceutical equivalents that produce the same effects in patients.

CHOOSE THE BEST ANSWER

Answers are at the end of the book.

1. ____________________ are drugs that activate receptors to accelerate or slow normal cellular function.
   a. Channels
   b. Agonists
   c. Antagonists
   d. Protein binding

2. The time between the onset of action and the time when the MEC is reached by declining blood concentrations is the
   a. duration of action.
   b. MTC.
   c. therapeutic window.
   d. dose-response curve.

3. When studying concentration and effect, __________________ is the time MEC is reached and the response occurs.
   a. therapeutic window
   b. MTC
   c. onset of action
   d. blood concentration

4. Most absorption of oral drugs occurs in the
   a. stomach.
   b. small intestine.
   c. large intestine.
   d. liver

5. Complex proteins in the liver that catalyze chemical reactions are
   a. enzymes.
   b. metabolites.
   c. antagonists.
   d. nephrons.

6. ________________________ refers to the transfer of drug into the blood from an administered drug product.
   a. Absorption
   b. Excretion
   c. Distribution
   d. Metabolism

7. "First-pass metabolism" occurs at the
   a. stomach.
   b. kidney.
   c. liver.
   d. small intestine.

8. The main functional unit of the kidney is the/an
   a. glomerulus.
   b. enzyme.
   c. metabolite.
   d. nephron.

9. The body's process of transforming drugs is called
   a. distribution.
   b. metabolism.
   c. excretion.
   d. absorption.

10. When bases dissociate, they become
    a. biological.
    b. acids.
    c. ionized.
    d. unionized.

11. The ____________________ filter(s) the blood and remove waste materials from it.
    a. kidneys
    b. liver
    c. gallbladder
    d. small intestine

12. The blood-filtering process of the nephron is called
    a. absorption.
    b. metabolism.
    c. glomerular filtration.
    d. excretion.

13. The bioavailability of a drug product compared to the same drug in a rapidly administered IV is
    a. absolute.
    b. relative.
    c. induction.
    d. inhibition.

14. Drug products that contain the same active ingredients in the same dosage form but different salt forms are
    a. pharmaceutical equivalents.
    b. bioequivalent drug products.
    c. pharmaceutical alternatives.
    d. antagonist drug products.

STUDY NOTES

Use this area to write important points you'd like to remember.

— 12 —

FACTORS AFFECTING DRUG ACTIVITY

KEY CONCEPTS

Test your knowledge by covering the information in the right-hand column.

human variability — Differences in age, weight, genetics, and gender are among the significant factors that influence the differences in medication responses among people.

age — Drug distribution, metabolism, and excretion are quite different in the neonate and infant than in adults because their organ systems are not fully developed. Children metabolize certain drugs more rapidly than adults. The elderly typically consume more drugs than other age groups. They also experience physiological changes that significantly affect drug action.

pregnancy — A number of physiological changes that occur in women in the latter stages of pregnancy tend to reduce the rate of absorption.

genetics — Genetic differences can cause differences in the types and amounts of proteins produced in the body, which can result in differences in drug action.

pharmacogenomics — A new field of study that defines the hereditary basis of individual differences.

body weight — Weight adjustments may be needed for individuals whose weight is more than 50% higher than the average adult weight. Weight adjustments are also made for children, or unusually small, emaciated, or obese adult patients.

common adverse reactions — Anorexia, nausea, vomiting, constipation, and diarrhea are among the most common adverse reactions to drugs.

allergic reactions — Almost any drug, in almost any dose, can produce an allergic or hypersensitive reaction in a patient. Anaphylactic shock is a potentially fatal hypersensitivity reaction.

teratogenicity
The ability of a substance to cause abnormal fetal development when given to pregnant women.

disease states
The disposition and effect of some drugs can be altered in one person but not in another by the presence of diseases other than the one for which a drug is used. Hepatic, cardiovascular, renal, and endocrine disease all increase the variability in drug response. For example, decreased blood flow from cardiovascular disorders can delay or cause erratic drug absorption.

drug–drug interactions
These can result in either increases or decreases in therapeutic effects or adverse effects.

displacement
Displacement of one drug from protein binding sites by a second drug increases the effects of the displaced drug. Decreased intestinal absorption can occur when orally taken drugs combine to produce nonabsorbable compounds, e.g., when magnesium hydroxide and oral tetracycline bind.

enzyme induction
Caused when drugs activate metabolizing enzymes in the liver, increasing the metabolism of other drugs affected by the same enzymes.

enzyme inhibition
When a drug blocks the activity of metabolic enzymes in the liver.

urinary reabsorption
Some drugs raise urinary pH, lessening renal reabsorption, e.g., sodium bicarbonate raises pH and will cause increased elimination of phenobarbital.

additive effects
Occur when two drugs with similar pharmacological actions are taken, e.g., alcohol and a sedative together produce increased sedation.

synergism
Occurs when two drugs with different sites or mechanisms of action produce greater effects than the sum of individual effects, e.g., acetaminophen and aspirin together produce increased anticoagulation.

drug–diet interactions
The physical presence of food in the gastrointestinal tract can alter absorption by interacting chemically (e.g., certain medications and tetracycline); improving the water-solubility of some drugs by increasing bile secretion; affecting the performance of the dosage form (e.g., altering the release characteristics of polymer-coated tablets); altering gastric emptying; altering intestinal movement; altering liver blood flow. Some foods contain substances that react with certain drugs, e.g., foods containing tyramine can react with monoamine oxidase (MAO) inhibitors.

FILL IN THE KEY TERM

Answers are at the end of the book.

adverse drug reaction
anaphylactic shock
antidote
carcinogenicity
cirrhosis
complexation
displacement
enzyme inhibition
hepatotoxicity
hypersensitivity
hyperthyroidism
hypothyroidism
idiosyncrasy
nephrotoxicity
obstructive jaundice
pharmacogenomics
potentiation
teratogenicity

1. ____________________ : An abnormal sensitivity generally resulting in an allergic reaction.
2. ____________________ : A potentially fatal hypersensitivity reaction that produces severe respiratory distress and cardiovascular collapse.
3. ____________________ : An unexpected reaction the first time a drug is taken, generally due to genetic causes.
4. ____________________ : Toxicity of the liver.
5. ____________________ : The ability of a substance to harm the kidneys.
6. ____________________ : The ability of a substance to cause cancer.
7. ____________________ : The ability of a substance to cause abnormal fetal development when given to pregnant women.
8. ____________________ : When one drug with no inherent activity of its own increases the activity of another drug, producing an effect.
9. ____________________ : Field of study that defines the hereditary basis of individual differences in ADME processes.
10. ____________________ : When one drug blocks the activity of metabolic enzymes in the liver.
11. ____________________ : When one drug is moved from protein binding sites by a second drug, resulting in increased effects of the displaced drug.
12. ____________________ : Obstruction of bile duct causing accumulations in the liver.
13. ____________________ : A drug that antagonizes the toxic effect of another.
14. ____________________ : A condition in which thyroid hormone secretions are below normal, often referred to as an underactive thyroid.

15. ____________________ : A condition in which thyroid hormone secretions are above normal, often referred to as an overactive thyroid.

16. ____________________ : When two different molecules associate or attach to each other.

17. ____________________ : A chronic and potentially fatal liver disease causing loss of function and resistance to blood flow through the liver.

18. ____________________ : An unintended side effect that is negative.

TRUE/FALSE

Indicate whether the statement is true or false in the blank. Answers are at the end of the book.

_____ 1. Physiological changes in pregnancy tend to increase the rate of absorption of drugs.

_____ 2. Adults experience a decrease in many physical functions between the ages of 30 and 70 years.

_____ 3. Lower cardiac output in the elderly tends to slow the distribution of drugs.

_____ 4. Foods containing tyramine, such as aged cheeses, may produce dangerous interactions with some medications.

_____ 5. Ibuprofen can cause nephrotoxicity.

_____ 6. The activity of metabolizing enzymes in the liver is reduced in hypothyroidism.

_____ 7. Some anticancer drugs are considered carcinogenic.

_____ 8. Iron salts form nonabsorbable complexes with tetracycline.

_____ 9. Phenobarbital is an example of an enzyme inducer.

_____ 10. Eating too much spinach can cause problems for patients who are taking some types of anticoagulants.

EXPLAIN WHY

Explain why these statements are true or important. Check your answers in the text. Discuss any questions you may have with your instructor.

1. Give at least three reasons why drug–drug interactions can increase the effects of drugs.

2. Give at least three reasons why drug–drug interactions can decrease the effects of drugs.

3. Give at least three reasons diet can affect drug activity.

IN THE WORKPLACE ACTIVITIES

1. Working in pairs, read about pregnancy and medicines at this FDA website: www.fda.gov/ForConsumers/ByAudience/ForWomen/ucm118567.htm. Discuss with your partner what precautions are needed for using medicines by pregnant women.
2. Working in pairs, read about reducing sodium intake at this FDA website: www.fda.gov/for-consumers/consumerupdates/ucm327369.htm Discuss with your partner the different disease states where sodium intake should be restricted. Also discuss the practical implications of reducing sodium intake.

CHOOSE THE BEST ANSWER

Answers are at the end of the book.

1. A potentially fatal allergic reaction is called
   a. nephrotoxicity.
   b. idiosyncrasy.
   c. teratogenicity.
   d. anaphylactic shock.

2. Placebo effects can be due to
   a. pregnancy.
   b. psychological factors.
   c. body weight.
   d. gender.

3. _________ has decreased elimination in cirrhosis.
   a. Atenolol
   b. Digoxin.
   c. Metoprolol
   d. Ranitidine

4. Hypersensitivity generally happens because a patient develops
   a. anaphylaxis.
   b. anorexia.
   c. antibodies.
   d. hepatotoxicity.

5. Hepatotoxicity is associated with the
   a. central nervous system.
   b. liver.
   c. kidneys.
   d. small intestine.

6. Unpleasant physical symptoms when some drugs are discontinued is called
   a. psychological dependence.
   b. idiosyncrasy.
   c. anaphylaxis.
   d. physiological dependence.

7. Antineoplastics can cause bone marrow suppression; and this is an example of
   a. hematological effects.
   b. carcinogenicity.
   c. teratogenicity.
   d. nephrotoxicity.

8. The ability of a substance to cause abnormal fetal development when given to pregnant women is called
   a. hematological effects.
   b. idiosyncrasy.
   c. nephrotoxicity.
   d. teratogenicity.

9. _________________ occurs when two drugs with similar pharmacological effects produce greater effects than the sum of individual effects when taken together.
   a. Synergism
   b. Additive effects
   c. Interference
   d. Displacement

10. Drugs that increase activity of metabolizing enzymes in the liver cause
    a. glomerular filtration.
    b. enzyme induction.
    c. renal secretion.
    d. enzyme inhibition.

11. A drug given to block or reduce toxic effects of another drug is called a(n)
    a. antidote.
    b. synergist.
    c. inducer.
    d. MAO inhibitor.

12. Increased protein in the diet appears to
    a. decrease glomerular filtration.
    b. increase metabolism.
    c. decrease metabolism.
    d. have no effect on metabolism.

13. Eating aged cheeses is especially dangerous when taking
    a. aspirin.
    b. phenelzine.
    c. acetaminophen.
    d. furosemide.

STUDY NOTES

Use this area to write important points you'd like to remember.

– 13 –

COMMON DRUGS & THEIR USES

CLASSIFICATION OF DRUGS

There are thousands of drugs used in pharmacy. A basic familiarity with these drugs and their uses will enhance your skills as a pharmacy technician.

Drugs may be classified in different ways. One way is *based on their main therapeutic indication or action.* An example of such a group name would be **antibiotic,** which describes drugs that work by destroying pathogenic organisms (microorganisms that cause disease). Another example of a group name would be **analgesic,** which describes drugs that are used in the alleviation of pain. Following is a sample list of such group names:

Analgesics
Anesthetic Agents
Anti-infectives
Antineoplastics
Cardiovascular Agents
Dermatologicals
Electrolytic Agents
Gastrointestinal & Urinary Tract Agents
Hematological Agents
Hormones & Modifiers
Immunobiologic Agents
Musculoskeletal Agents
Neurological Agents
Ophthalmic & Otic Agents
Psychotropic Agents
Respiratory Agents

Another arrangement of drugs is by *specific classification based on how the drug actually works.* Drugs classified this way generally share these characteristics:

1. similar chemical structure
2. similar mechanism of action
3. similar effects (including side effects)

An example of such a classification is the **cephalosporins,** which is found in the antibiotics group. Drugs in the cephalosporins classification share the above mentioned characteristics with each other but not necessarily with other antibiotics, which may have different characteristics. Another example of such a group would be **xanthine derivatives,** which are found in the bronchodilators group.

Example:

Trade Name	Generic Name	Group	Classification
Keflex	Cephalexin	Antibiotic	Cephalosporin
Theo-Dur	Theophylline	Bronchodilator	Xanthine Derivative

Below is a sample list of such classifications:

Angiotensin Converting Enzyme inhibitors (ACE inhibitors)
α-adrenergic agonists
β-blockers
Calcium channel blockers
Cephalosporins
Corticosteroids
Histamine_2 blockers
Loop Diuretics

For national certification exams, you will need to know the main groups of drugs used in the retail and hospital setting. More specific classifications are not emphasized. You will learn most of this information on the job: as you handle these medications over and over again, you'll see which drugs are the most commonly used and which are not.

STUDY TIP — DRUG CARDS

A good way to help you remember the most common drugs used in pharmacy is to create drug cards. Drug cards are easy to make by writing information about drugs on small index cards. You can then use these to study until you have memorized the information.

Information needed on your drug card:

1. Trade name of drug

2. Generic name of drug

3. Classification

KEY CONCEPTS

Test your knowledge by covering the information in the right-hand column.

USAN	The United States Adopted Names Council (USAN) designates nonproprietary names for drugs.
drug classes	Group names for drugs that have similar activities or are used for the same type of diseases and disorders.
stems	Common stems or syllables that are used to identify the different drug classes and in making new nonproprietary names. They are approved and recommended by the USAN.
neurotransmitter	Substances that carry the impulses from one neuron to another.
blocker	Another name for an antagonist drug—because antagonists block the action of a neurotransmitter.
homeostasis	The state of equilibrium of the body.
mimetic	Another term for an agonist drug—because agonists imitate, or "mimic," the action of the neurotransmitter.
analgesia	A state in which pain is not felt even though a painful condition exists.
antipyretic	Reduces fever.
opiate-type analgesics	Drugs related to morphine and codeine that can be habit forming and are used for pain relief.
narcotic analgesics	Same as opiate-type analgesics.
salicylates	Drugs related to aspirin that are used to relieve mild to moderate pain, and have anti-inflammatory and antipyretic properties.
NSAIDs	Nonsteroidal, anti-inflammatory drugs that also have antipyretic and analgesic properties.
acetaminophen	A drug that relieves mild to moderate pain and has antipyretic properties.
local anesthetics	Drugs that block pain conduction from the peripheral nerves to the central nervous system without causing a loss of consciousness.
surgical anesthesia	The stage of anesthesia in which surgery can be safely conducted.
medullary paralysis	An overdose of anesthesia that paralyzes the respiratory and heart centers of the medulla, leading to death.
antibiotic (antimicrobial)	Drug that destroys microorganisms.
antiviral	Drug that attacks a virus.

antifungal	Drug that destroys fungi or inhibits growth of fungi.
antimycobacterial	Drug that attacks the organisms that cause tuberculosis and leprosy.
antiprotozoal	Drug that destroys protozoa.
anthelmintic	Drug that destroys worms.
bactericidal	Bacteria killing.
bacteriostatic	Bacteria inhibiting.
virustatic	Drug that inhibits the growth of viruses.
antineoplastic	Drug that inhibits new growth of cancer cells.
lymphocyte	A type of white blood cell that releases antibodies that destroy disease cells.
metastasis	When cancer cells spread beyond their original site.
neoplasm	A new and abnormal tissue growth, often referring to cancer cells.
remission	A state in which cancer cells are inactive.
arrhythmia	An abnormal heart rhythm.
cardiac cycle	The contraction and relaxation of the heart that pumps blood through the cardiovascular system.
diastolic pressure	The minimum blood pressure when the heart relaxes; the second number in a blood pressure reading.
electrocardiogram (EKG or ECG)	A graph of the heart's rhythm.
embolism	A clot that has traveled in the bloodstream to a point where it obstructs flow.
myocardium	Heart muscle.
systolic pressure	The maximum blood pressure when the heart contracts; the first number in a blood pressure reading.
thrombus	A blood clot.
antianginals	Drugs used to treat cardiac-related chest pain (angina).
antiarrhythmics	Drugs used to treat irregular heart rhythms.
antihypertensives	Drugs used to reduce a sustained elevation in blood pressure.
vasopressors	Drugs used to increase blood pressure.
antihyperlipidemics	Drugs used to lower high levels of cholesterol.

KEY CONCEPTS

Test your knowledge by covering the information in the right-hand column.

thrombolytics	Drugs used to dissolve blood clots.
anticoagulants	Drugs used to prevent blood clot formation.
beta blockers	Drugs that reduce the oxygen demands of the heart muscle.
calcium channel blockers	Drugs that relax the heart by reducing heart conduction.
diuretics	Drugs that decrease blood pressure by decreasing blood volume.
ACE inhibitors	The "pril" drugs that relax the blood vessels.
vasodilators	Drugs that relax and expand the blood vessels.
dermatological	A product that is used to treat a skin condition.
integumentary system	The skin.
anion	A negatively charged ion.
cation	A positively charged ion.
dissociation	When a compound breaks down and separates into smaller components.
electrolytes	A substance that in solution forms ions that conduct an electrical current.
extracellular fluids	The fluid outside the body's individual cells found in plasma and tissue fluid.
intracellular fluids	Cell fluid.
interstitial fluid	Tissue fluid.
ions	Electrically charged particles.
vaccine	A suspension containing infectious agents used to boost the body's immune system response.
chyme	The semiliquid form of food as it enters the intestinal tract.
peristalsis	The wavelike motion of the intestines that moves food through them.
enzymes	Substances in the body that help the body to break down molecules.
antidiarrheals	Drugs used to treat diarrhea.
antiemetics	Drugs used to treat nausea and vomiting.
antacids	Drugs used to neutralize acid.

laxatives	Drugs that promote defecation.
stool softeners	Drugs that promote mixing of fatty and watery internal substances to soften the stool's contents and ease the evacuation of feces.
hematological agents	Drugs that affect the blood.
fibrin	The fiber that serves as the structure for clot formation.
anemia	A decrease in hemoglobin or red blood cells.
hemostatic drugs	Drugs that prevent excessive bleeding.
hormones	Chemicals that are secreted in the body by the endocrine system's ductless glands.
corticosteroid	Hormonal steroid substances produced by the cortex of the adrenal gland.
endocrine system	The system of hormone secreting glands.
hyperthyroidism	Overproduction of thyroid hormone.
hypothyroidism	Underproduction of thyroid hormone.
insulin	A hormone that controls the body's use of glucose.
glucagon	A hormone that helps convert amino acid to glucose.
diabetes mellitus	A condition in which the body does not produce enough insulin or is unable to use insulin efficiently.
serum glucose	Blood sugar.
androgens	Male sex hormones.
estrogen	Female sex characteristic hormone that is involved in calcium and phosphorus conservation.
progesterone	Female sex characteristic hormone that is involved in ovulation prevention.
testosterone	The primary androgen (male sex hormone).
gout	A painful inflammatory condition in which excess uric acid accumulates in the joints.
rheumatoid arthritis	A chronic and often progressive inflammatory condition with symptoms that include swelling, feelings of warmth, and joint pain.
osteoarthritis	A disorder characterized by weight-bearing bone deterioration, decreasing range of motion, pain, and deformity.
Parkinson's disease	A progressive neuromuscular condition.
Alzheimer's disease	A progressive dementia condition.

KEY CONCEPTS

Test your knowledge by covering the information in the right-hand column.

epilepsy	A neurologic disorder characterized by seizures.
migraine headaches	A type of headache associated with possible auras and pain.
ophthalmic agents	Drugs used to treat conditions of the eye.
conjunctivitis	Inflammation of the eyelid lining.
glaucoma	A disorder characterized by high pressure within the eye.
mydriatics	Drugs that dilate the pupil.
sedatives	Drugs that are intended to relax and calm.
hypnotics	Drugs that are intended to induce sleep.
bipolar disorder	A disorder characterized by mood swings.
depression	A disorder characterized by low mood.
asthma	A condition characterized by chronic airway inflammation.
emphysema	A condition associated with chronic airway obstruction.
antihistamines	Drugs that replace histamine at histamine receptor sites.
decongestants	Drugs that cause mucous membrane vasoconstriction.
antitussives	Drugs that are used to treat coughs.
bronchodilators	Drugs that are used to relieve bronchospasm.

STUDY NOTES

Use this area to write important points you'd like to remember.

FILL IN THE BLANKS

Match the drug classifications with the generic drug names. Answers are at the end of the book.

Analgesic
Anesthetic
Antidiabetic
Anti-infective
Antineoplastic
Cardiovascular Agent
Dermatological Agent
Electrolytic Agent
Gastrointestinal Agent
Musculoskeletal Agent
Psychotropic Agent
Respiratory Agent

Generic Name	Drug Classification
1. Albuterol	______________________
2. Alprazolam	______________________
3. Amlodipine	______________________
4. Amoxicillin/Clavulanate	______________________
5. Ampicillin	______________________
6. Atenolol	______________________
7. Atorvastatin	______________________
8. Azithromycin	______________________
9. Carisoprodol	______________________
10. Carvedilol	______________________
11. Cefaclor	______________________
12. Cetirizine	______________________
13. Ciprofloxacin	______________________
14. Citalopram	______________________
15. Clarithromycin	______________________
16. Clopidogrel	______________________
17. Clotrimazole/Betamethasone	______________________
18. Diltiazem	______________________
19. Docusate sodium	______________________
20. Doxazosin	______________________
21. Escitalopram	______________________
22. Fluconazole	______________________

23. Fluticasone ____________________
24. Furosemide ____________________
25. Glyburide ____________________
26. Heparin ____________________
27. Hydrocortisone cream ____________________
28. Ibuprofen ____________________
29. Insulin ____________________
30. Lactulose ____________________
31. Levofloxacin ____________________
32. Lisinopril ____________________
33. Loperamide ____________________
34. Metoprolol ____________________
35. Nifedipine ____________________
36. Nitroglycerin ____________________
37. Omeprazole ____________________
38. Pantoprazole ____________________
39. Potassium chloride ____________________
40. Procaine ____________________
41. Sertraline ____________________
42. Silver sulfadiazine ____________________
43. Sodium chloride ____________________
44. Tamoxifen ____________________
45. Tramadol ____________________
46. Trimethobenzamide ____________________
47. Trimethoprim/Sulfamethoxazole ____________________
48. Warfarin ____________________

MATCH THE GENERIC AND BRAND NAMES

In the following exercises, match each brand name with its generic name. The list of the Top 200 Most-Prescribed Drugs by Classification (Appendix A) should be helpful. Answers are at the end of the book.

ANALGESICS

Generic Name	Brand Name
1. Fentanyl _____	a. Celebrex
2. Tramadol _____	b. Tylenol with Codeine
3. Celecoxib _____	c. Motrin
4. Diclofenac _____	d. Ultram
5. buprofen _____	e. Percocet
6. Meloxicam _____	f. Vicodin
7. Naproxen _____	g. MS Contin
8. Acetaminophen, codeine _____	h. Duragesic
9. Hydrocodone, acetaminophen _____	i. Mobic
10. Morphine sulfate ER _____	j. Naprosyn
11. Oxycodone _____	k. Oxycontin
12. Oxycodone, acetaminophen _____	l. Voltaren

ANTI-INFECTIVES

Generic Name	Brand Name
1. Amoxicillin _____	a. Flagyl
2. Amoxicillin, clavulanate _____	b. Cipro
3. Azithromycin _____	c. Augmentin
4. Cephalexin _____	d. Valtrex
5. Ciprofloxacin _____	e. Nizoral
6. Clindamycin HCl _____	f. Nilstat
7. Doxycycline Hyclate _____	g. Amoxil
8. Levofloxacin _____	h. Zithromax
9. Metronidazole _____	i. Tamiflu
10. Minocycline HCl _____	j. Bactrim
11. Mupirocin _____	k. Bactroban

12. Nitrofurantoin mono-macro _____
13. Sulfamethoxazole, trimethoprim _____
14. Fluconazole _____
15. Ketoconazole _____
16. Nystatin _____
17. Acyclovir _____
18. Oseltamivir _____
19. Valacyclovir _____

l. Keflex
m. Cleocin
n. Minocin
o. Doryx
p. Macrobid
q. Levaquin
r. Diflucan
s. Zovirax

CARDIOVASCULAR AGENTS I

Generic Name

1. Carvedilol _____
2. Clopidogrel _____
3. Warfarin _____
4. Atorvastatin _____
5. Ezetimibe _____
6. Fenofibrate _____
7. Gemfibrozil _____
8. Lovastatin _____
9. Pravastatin _____
10. Rosuvastatin _____
11. Simvastatin _____

Brand Name

a. Plavix
b. Lipitor
c. Coreg
d. Coumadin
e. Lopid
f. Crestor
g. Zetia
h. Mevacor
i. Zocor
j. Pravachol
k. Tricor

CARDIOVASCULAR AGENTS II

Generic Name

1. Nevibolol _____
2. Amlodipine _____
3. Amlodipine, benazepril _____
4. Atenolol _____
5. Benazepril _____
6. Clonidine _____

Brand Name

a. Bystolic
b. Catapres
c. Lotrel
d. Lopressor
e. Zestoretic
f. Hyzaar

MATCH THE GENERIC AND BRAND NAMES (CONT'D)

Answers are at the end of the book.

CARDIOVASCULAR AGENTS II (CONT'D)

Generic Name	Brand Name
7. Diltiazem _____	g. Benicar
8. Enalapril _____	h. Norvasc
9. Furosemide _____	i. Tenormin
10. Hydrochlorothiazide _____	j. Lotensin
11. Lisinopril _____	k. Lasix
12. Lisinopril, hydrochlorothiazide _____	l. Calan
13. Losartan _____	m. Xarelto
14. Losartan, hydrochlorothiazide _____	n. Cardizem
15. Metoprolol succinate _____	o. Cozaar
16. Metoprolol tartrate _____	p. Diovan HCT
17. Nifedipine ER _____	q. Dyazide
18. Olmesartan _____	r. Inderal
19. Propranolol _____	s. Vasotec
20. Ramipril _____	t. Microzide
21. Spironolactone _____	u. Toprol
22. Triamterene, hydrochlorothiazide _____	v. Zestril
23. Valsartan _____	w. Procardia XL
24. Valsartan, hydrochlorothiazide _____	x. Altace
25. Verapamil _____	y. Aldactone
26. Rivaroxaban _____	z. Diovan

DERMATOLOGICALS

Generic Name	Brand Name
1. Clindamycin phosphate _____	a. Temovate
2. Clobetasol _____	b. Vistaril
3. Clotrimazole, betamethasone _____	c. Cleocin
4. Hydroxyzine pamoate _____	d. Lotrisone

Gastrointestinal

Generic Name

1. Dicyclomine _____
2. Phentermine HCl _____
3. Dexlansoprazole _____
4. Esomeprazole _____
5. Famotidine _____
6. Lansoprazole _____
7. Omeprazole i
8. Pantoprazole _____
9. Ranitidine l
10. Ranitidine h
11. Ondansetron e
12. Polyethylene glycol d

Brand Name

a. Dexilant
b. Pepcid
c. Adipex-P
d. Miralax
e. Zofran
f. Bentyl
g. Nexium
h. Zantac
i. Prilosec
j. Prevacid
k. Protonix
l. Zantac

Hormones & Modifiers I

Generic Name

1. Insulin lispro b
2. Hydrocortisone g
3. Methylprednisolone d
4. Prednisolone e
5. Prednisone h
6. Ethinyl estradiol, etonogestrel f
7. Conjugated estrogens b
8. Estradiol i
9. Insulin glargine a

Brand Name

a. Lantus
b. Premarin
c. Humalog
d. Medrol
e. Sterapred
f. Nuvaring
g. Cortef
h. Orapred
i. Estrace

MATCH THE GENERIC AND BRAND NAMES (CONT'D)

Answers are at the end of the book.

HORMONES & MODIFIERS II

Generic Name	Brand Name
1. Glimepiride _____	a. DiaBeta
2. Glipizide _____	b. Amaryl
3. Glyburide _____	c. Actos
4. Metformin _____	d. Januvia
5. Pioglitazone _____	e. Glucotrol
6. Sitagliptin _____	f. Glucophage

HORMONES & MODIFIERS III

Generic Name	Brand Name
1. Ethinyl estradiol, norgestimate _____	a. Tri-Sprintec
2. Ethinyl estradiol, norethindrone, iron _____	b. Viagra
3. Ethinyl estradiol, norgestimate _____	c. Microgestin Fe
4. Sildenafil _____	d. Sprintec
5. Tadalafil _____	e. Provera
6. Medroxyprogesterone acetate _____	f. Synthroid
7. Desiccated thyroid _____	g. Cialis
8. Levothyroxine _____	h. Armour Thyroid

MUSCULOSKELETAL

Generic Name	Brand Name
1. Allopurinol _____	a. Fosamax
2. Hydroxychloroquine _____	b. Robaxin
3. Methotrexate _____	c. Zyloprim
4. Carisoprodol _____	d. Plaquenil
5. Cyclobenzaprine _____	e. Zanaflex
6. Methocarbamol _____	f. Rheumatrex
7. Tizanidine _____	g. Flexeril
8. Alendronate _____	h. Soma

NEUROLOGIC

Generic Name	Brand Name
1. Butalbital, acetaminophen, caffeine _____	a. Antivert
2. Meclizine _____	b Fioricet
3. Sumatriptan _____	c. Topamax
4. Ropinirole _____	d. Keppra
5. Lyrica _____	e. Requip
6. Gabapentin __i__	f. regabalin
7. Lamotrigine _____	g. Imitrex
8. Levetiracetam __d__	h. Lamictal
9. Topiramate _____	i. Neurontin

PSYCHOTROPIC I

Generic Name	Brand Name
1. Alprazolam _____	a. Aricept
2. Buprenorphine, naloxone _____	b. Valium
3. Citalopram _____	c. Xanax
4. Clonazepam _____	d. Suboxone
5. Diazepam _____	e. Ativan
6. Donepezil HCl ______	f. Celexa
7. Duloxetine _____	g. Seroquel
8. Lorazepam _____	h. Klonopin
9. Quetiapine _____	i. Cymbalta

PSYCHOTROPIC II

Generic Name	Brand Name
1. Escitalopram _____	a. Lexapro
2. Fluoxetine _____	b. Prozac
3. Mirtazapine _____	c. Desyrel
4. Paroxetine _____	d. Remeron

MATCH THE GENERIC AND BRAND NAMES (CONT'D)

Answers are at the end of the book.

PSYCHOTROPIC II (CONT'D)

Generic Name	Brand Name
5. Sertraline _____	e. Zoloft
6. Bupropion _____	f. Ambien
7. Trazodone _____	g. Paxil
8. Temazepam _____	h. Wellbutrin
9. Zolpidem _____	i. Restoril

PSYCHOTROPIC III

Generic Name	Brand Name
1. Methylphenidate HCl _____	a. Adderall
2. Amphetamine, dextroamphetamine _____	b. Risperdal
3. Lisdexamphetamine _____	c. Abilify
4. Aripiprazole _____	d. Vyvanse
5. Risperidone _____	e. Ritalin

RESPIRATORY I

Generic Name	Brand Name
1. Beclomethasone _____	a. Spiriva
2. Budesonide, formoterol _____	b. Singulair
3. Fluticasone _____	c. Qvar
4. Fluticasone, salmeterol _____	d. Symbicort
5. Mometasone _____	e. Advair Diskus
6. Montelukast _____	f. Flovent
7. Tiotropium _____	g. Nasonex

RESPIRATORY II

Generic Name	Brand Name
1. Cetirizine _____	a. Phenergan
2. Levocetirizine _____	b. Proair HFA
3. Loratadine _____	c. Zyrtec
4. Promethazine _____	d. Claritin

RESPIRATORY II (CONT'D)

Generic Name	Brand Name
5. Benzonatate _____	e. Tessalon
6. Albuterol _____	f. Xyzal

URINARY

Generic Name	Brand Name
1. Oxybutynin _____	a. Cardura
2. Doxazosin _____	b. Proscar
3. Finasteride _____	c. Flomax
4. Tamsulosin _____	d. Ditropan

IN THE WORKPLACE ACTIVITIES

1. Go to the website for Vaccine Information Sheets at http://immunize.org/vis/. Obtain the VIS sheets for the following immunizations and record the date of the most recent revision: MMR, Hib, Hepatitis A, Hepatitis B, HPV-Gardasil 9, Shingles, Td, Influenza, Tdap.
2. Identify 10 high-alert medications and put flags on the products and the shelves where they are located.
3. Working in pairs, practice pronouncing the following drug names in front of your partner. The partner should provide a critique of the pronunciation of the drug names: hydrochlorothiazide, furosemide, enalapril, ibuprofen, atenolol, propranolol, Mevacor. Use the National Library of Medicine online resource MedlinePlus (https://www.nlm.nih.gov/medlineplus/druginformation.html) as a guide for pronunciation.

TRUE/FALSE

Indicate whether the statement is true or false in the blank. Answers are at the end of the book.

_____ 1. Once a suggested nonproprietary name for a drug is officially approved, it becomes the generic name of the drug.

_____ 2. Meperidene is a naturally occurring opiate.

_____ 3. Medullary paralysis is associated with an overdose of analgesic.

_____ 4. Ringworm is a type of fungal infection.

_____ 5. Vasopressors are used to treat hypertension.

_____ 6. Some antacids have significant drug interactions with tetracycline.

_____ 7. Hematopoietic drugs are used to treat excessive bleeding.

_____ 8. Insulin is a cure for diabetes.

_____ 9. Calcium channel blockers are sometimes used to prevent migraine headaches.

_____ 10. Meprobamate is a barbiturate.

EXPLAIN WHY

Explain why these statements are true or important. Check your answers in the text. Discuss any questions you may have with your instructor.

1. Why would an antibiotic not be appropriate for treating viral infections?

2. Why is it important for pharmacy technicians to be careful when handling antineoplastic drugs?

3. Why is it important for patients with diabetes to use a glucometer?

CHOOSE THE BEST ANSWER

Answers are at the end of the book.

1. The ________ designates nonproprietary names for drugs.
   a. manufacturer
   b. FDA
   c. USAN
   d. DEA

2. ________________ is a neurotransmitter.
   a. Estrogen
   b. Testosterone
   c. Insulin
   d. Epinephrine

3. A common, naturally occurring opiate-type drug is
   a. codeine.
   b. cocaine.
   c. meperidine.
   d. propoxyphene.

4. An inhalation anesthetic is
   a. propofol.
   b. isoflurane.
   c. etomidate.
   d. methohexital.

5. An antifungal drug is
   a. metronidazole.
   b. nystatin.
   c. indinavir.
   d. tetracycline.

6. The term used to denote the presence of a life-threatening cancerous group of cells or tumor is
   a. malignant.
   b. remission.
   c. benign.
   d. viral.

7. Drugs that decrease blood pressure by decreasing blood volume are called
   a. beta blockers.
   b. calcium channel blockers.
   c. diuretics.
   d. vasodilators.

8. Drugs that act to increase blood pressure are
   a. vasodilators.
   b. antihypertensives.
   c. vasopressors.
   d. ACE inhibitors.

9. A type of skin cancer:
   a. basal cells.
   b. keratoses.
   c. dandruff.
   d. cellulitus

10. The histamine receptor antagonist with the most drug interactions is
    a. ranitidine.
    b. cimetidine.
    c. famotidine.
    d. nizatidine.

11. A commonly ordered stool softener is
    a. docusate sodium.
    b. lactulose.
    c. bismuth subsalicylate.
    d. loperamide.

12. ______________ is used for vitamin B12 deficiency.
    a. Vitamin K
    b. Cyanocobalamin
    c. Ferrous sulfate
    d. Plasminogen

13. A posterior lobe hormone of the pituitary gland:
    a. oxytocin.
    b. thyroid.
    c. insulin.
    d. glucagon.

14. A drug that reduces uric acid and is used to treat gout is
    a. celecoxib.
    b. ibuprofen.
    c. allopurinol.
    d. cyclobenzaprine.

15. A drug commonly used to treat epilepsy is
    a. phenobarbital.
    b. sumatriptan.
    c. tacrine.
    d. donepezil.

— 14 —

INVENTORY MANAGEMENT

KEY CONCEPTS

Test your knowledge by covering the information in the right-hand column.

inventory	A list of goods or items a business uses in its normal operations.
open formulary	One that allows purchase of any medication that is prescribed.
closed formulary	A limited list of approved medications.
perpetual inventory system	A system that maintains a continuous record of every item in inventory so that it always shows the stock on hand.
turnover	The rate at which inventory is used.
fast mover	20% of the stock that accounts for 80% of the orders or prescriptions
reorder points	Maximum and minimum inventory levels for each drug.
point-of-sale (POS) system	A system in which the item is deducted from inventory as it is sold or dispensed.
automated dispensing system	A device that dispenses medications at point-of-use upon confirmation of an order communicated from a centralized computer system.
wholesalers	More than three-quarters of pharmaceutical manufacturers' sales are directly to drug wholesalers, who in turn resell their inventory to hospitals, pharmacies, and other pharmaceutical dispensers. They are government licensed and regulated.
340B Drug Pricing Program	A government plan that limits the cost that safety net providers such as Medicaid programs, federally qualified health centers, and qualified hospitals, pay for covered outpatient drugs.
drop shipments	Lower-volume, high-cost medications shipped on an as-needed basis directly from the manufacturer and billed through the wholesaler.
purchase order number	The number assigned to each order for identification.

storage Drugs must be stored according to manufacturer's specifications. Most drugs are kept in a fairly constant room temperature of 59°–86°F. The temperature of refrigeration should generally be 36°–46°F.

freshness Medications should be organized in a way that will dispense the oldest items first.

point-of-use stations In hospitals and other settings, medications are stocked in dispensing units throughout the facility that may be called supply stations or med stations.

Safety Data Sheets (SDSs) Instructions for hazardous substances such as chemotherapeutic agents that indicate when special handling and shipping is required.

reverse distributors Companies that specialize in returns to the manufacture of expired and discontinued drugs.

spoilage Inappropriate storage conditions or expired products automatically determine that a product is spoiled since in either case the chemical compounds in the drug product may have degraded.

controlled substances These substances are shipped separately and checked in by a pharmacist. A special order process must be used for Schedule II substances.

consignment stock High-cost medications that a pharmacy does not pay for until a patient purchases them.

durable and nondurable equipment supplies Durable equipment includes items such as walkers, wheelchairs, crutches, that are often bulky. Nondurable equipment includes items such as syringes, needles, and ostomy supplies.

unit dose package A package containing a single dose of medication.

Study Notes

Use this area to write important points you'd like to remember.

FILL IN THE KEY TERM

Answers are at the end of the book.

automated dispensing system
closed formulary
consignment stock
drop shipments
open formulary
perpetual inventory
point-of-use stations
purchase order number
reorder points
reverse distributors
SDSs
turnover
unit dose

1. ____________________: One that allows purchase of any medication that is prescribed.
2. ____________________: A limited list of approved medications.
3. ____________________: The rate at which inventory is used, generally expressed in number of days.
4. ____________________: Companies that specialize in returns to the manufacturer of expired and discontinue drugs.
5. ____________________: A system that maintains a continuous record of every item in inventory so that it always shows the stock on hand.
6. ____________________: Lower-volume, high-cost medications shipped on an as needed basis directly from the manufacturer and billed through the wholesaler.
7. ____________________: Minimum and maximum stock levels, which determine when a reorder is placed and for how much.
8. ____________________: Instructions for hazardous products.
9. ____________________: High-cost medications that the pharmacy does not pay for until the patient purchases them.
10. ____________________: A number assigned to each order for products that will allow it to be tracked and checked throughout the order process.
11. ____________________: A device that dispenses medications at point-of-use.
12. ____________________: A package containing a single dose of a medication.
13. ____________________: In hospitals and other settings, medications are stocked in units throughout the facility that may also be called supply stations or med stations.

TRUE/FALSE

Indicate whether the statement is true or false in the blank. Answers are at the end of the book.

_____ 1. The majority of pharmaceutical manufacturer sales are to wholesalers.

_____ 2. Medications should be organized in a way so the newest items will be dispensed first.

_____ 3. Reorder points are maximum and minimum inventory levels for a product.

_____ 4. Computerized ordering systems do not allow manual editing.

_____ 5. Point-of-sale systems adjust inventory as medications are sold or dispensed.

_____ 6. With computers keeping records, printed copies are not needed.

_____ 7. Certain hazardous substances may not be shipped by air.

_____ 8. A closed formulary allows any medication prescribed to be purchased.

_____ 9. Refrigeration means 50°–59°F.

_____ 10. Bar codes are used to quickly identify a product.

EXPLAIN WHY

Explain why these statements are true or important. Check your answers in the text. Discuss any questions you may have with your Instructor.

1. Why are wholesalers used?

2. Why is knowing the turnover rate of a product important?

3. Why are reorder points used?

4. Why is it important to make hard copy of computerized reports?

5. Why is it important to back up computer files?

IN THE WORKPLACE ACTIVITY

1. Print labels from http://labels.fda.gov/ for drugs in the prescriptions on pages 70–75 of this book and simulate placing those products in their proper storage areas.

CHOOSE THE BEST ANSWER

Answers are at the end of the book.

1. The list of medications that are approved for use in a health-care system is called a
   a. turnover.
   b. formulary.
   c. therapeutic equivalent.
   d. wholesaler.

2. Pharmaceutical equivalents that produce the same effects in patients are
   a. generic equivalents.
   b. always less expensive.
   c. always more expensive.
   d. therapeutically equivalent.

3. _____________ is an expression for the rate at which inventory is used and is generally expressed in number of days.
   a. Reciprocal
   b. Turnover
   c. POS
   d. Availability

4. A general rule is that _____% of stock accounts for _____% of prescriptions.
   a. 20/80
   b. 30/70
   c. 40/60
   d. 50/50

5. An inventory system in which the item is deducted from inventory as it is dispensed is called a (an)
   a. automated dispensing unit.
   b. point-of-use system.
   c. formulary.
   d. point of sale system (POS).

6. Pyxis Med/Supply Station is a good example of a (an)
   a. automated point-of-use storage system.
   b. automated dispensing cabinet.
   c. robotic dispensing machine.
   d. mobile robot.

7. Safety Data Sheets (SDS) are required by _____________ for hazardous substances and provide hazard, handling, clean-up, and first aid information.
   a. OSHA
   b. State Board of Pharmacy
   c. FDA
   d. DEA

8. When reconciling an order, controlled substances are shipped separately, and should be checked in by a(an)
   a. technician.
   b. pharmacy clerk.
   c. staff member different from the person who placed the order.
   d. intern.

9. The p.o. number identifies the
   a. post office.
   b. pharmacy.
   c. point-of-sale.
   d. purchase order.

10. Walkers, wheel chairs, crutches, and bedpans are examples of
    a. GPO.
    b. DME.
    c. DOT.
    d. EPA.

Study Notes

Use this area to write important points you'd like to remember.

– 15 –

FINANCIAL ISSUES

KEY CONCEPTS

Test your knowledge by covering the information in the right-hand column.

third-party programs Another party besides the patient or the pharmacy that pays for some or all of the cost of medication: essentially, an insurer.

pharmacy benefit manager A company that administers drug benefit programs for insurance companies, HMOs, and self-insured employers.

co-insurance Essentially an agreement between the insurer and the insured to share costs.

co-pay The portion of the cost of prescriptions that patients with third-party insurance must pay.

maximum allowable cost The amount paid by the insurer is not equal to the retail price normally charged, but is determined by a formula described in a contract between the insurer and the pharmacy. There is a maximum allowable cost (MAC) per tablet or other dispensing unit that an insurer or PBM will pay for a given product.

usual and customary (U&C) The MAC is often determined by survey of the usual and customary (U&C) prices for a prescription within a given geographic area. This is also referred to as the UCR (usual, customary, and reasonable) price for the prescription.

deductible A set amount that must be paid by the patient for each benefit period before the insurer will cover additional expenses.

prescription drug benefit Cards that contain necessary billing information for pharmacies, including the patient's identification number, group number, and co-pay amount.

HMO (health maintenance organization Health-care networks that usually do not cover expenses incurred outside the network and often require generic substitution.

point of service (POS) Health-care network where the patient's primary care physician must be a member and costs outside the network may be partially reimbursed.

preferred provider Health-care network that reimburses expenses outside the network at a lower rate than inside the network and usually requires generic substitution.

Medicare National health insurance for people over the age of 65, disabled people under the age of 65, and people with kidney failure.

Medicaid A federal-state program for the needy.

coordination of benefits Process to provide maximum coverage for health benefits when a patient has coverage by two plans.

workers' compensation Compensation for employees accidentally injured on the job.

online adjudication Most prescription claims are now filed electronically by online claim submission and online adjudication of claims. In online adjudication, the technician uses the computer to determine the exact coverage for each prescription with the appropriate third party.

dispensing code When brand name drugs are dispensed, numbers corresponding to the reason for submitting the claim with brand name drugs are entered in a dispensing code indicator field in the prescription system.

patient identification number The number assigned to the patient by the insurer that is indicated on the drug benefit card. If it does not match the code for the patient in the insurer's computer (with the same sex and other information) a claim may be rejected.

age limitations Many prescription drug plans have age limitations for children or dependents of the cardholder.

refills Most third-party plans require that most of the medication has been taken before the plan will cover a refill of the same medication.

maintenance medications Many managed care health programs require mail order pharmacies to fill prescriptions for maintenance medications.

rejected claims When a claim is rejected, the pharmacy technician can telephone the insurance plan's pharmacy help desk to determine if the patient is eligible for coverage.

FILL IN THE KEY TERM

Answers are at the end of the book.

CHAMPUS
CHAMPVA
co-insurance
co-pay
CMS-1500 form
coordination of benefits
CPT code
HMO
maximum allowable cost (MAC)
Medicaid
Medicare
online adjudication
patient assistance programs
pharmacy benefits managers
POS
PPO
TRICARE
U&C or UCR
workers' compensation

1. ____________________ : Companies that administer drug benefit programs.
2. ____________________ : The resolution of prescription coverage through the communication of the pharmacy computer with the third-party computer.
3. ____________________ : An agreement for cost-sharing between the insurer and the insured.
4. ____________________ : The portion of the price of medication that the patient is required to pay.
5. ____________________ : Civilian health and medical program of the uniformed services.
6. ____________________ : The maximum price per tablet (or other dispensing unit) an insurer or PBM will pay for a given product.
7. ____________________ : The maximum amount of payment for a given prescription, determined by the insurer to be a reasonable price.
8. ____________________ : A network of providers for which costs are covered inside but not outside of the network.
9. ____________________ : A network of providers where the patient's primary care physician must be a member and costs outside the network may be partially reimbursed.
10. ____________________ : A network of providers where costs outside the network may be partially reimbursed and the patient's primary care physician need not be a member.
11. ____________________ : Civilian health and medical program of the department of Veterans Affairs.
12. ____________________ : Provides health-care benefits for eligible uniform service members, retirees, and family members.
13. ____________________ : Identifiers used for billing pharmacist-provided MTM services..
14. ____________________ : A federal program providing health care to people with certain disabilities or who are age 65 and over.

15. ____________________ : A federal-state program, administered by the states, providing health care for the needy.

16. ____________________ : The standard form used by health-care providers to bill for services.

17. ____________________ : An employer compensation program for employees accidentally injured on the job.

18. ____________________ : Manufacturer sponsored prescription drug programs for the needy.

19. ____________________ : Process to provide maximum coverage when a patient has coverage by two plans.

TRUE/FALSE

Indicate whether the statement is true or false in the blank. Answers are at the end of the book.

_____ 1. The amount paid by a co-insurer to the pharmacy is equal to the wholesale price of a drug.

_____ 2. Medicare and Medicaid are examples of public health insurance programs.

_____ 3. A pharmacy benefits manager is a company that administers drug benefits programs.

_____ 4. Tier one drugs are usually generics.

_____ 5. Many third-party programs have drug formularies.

_____ 6. Prior authorization is a procedure to gain third-party coverage for a drug that is not automatically covered by a third party plan.

_____ 7. The CMS-1500 form is used to apply for an NPI.

_____ 8. An NPI number identifies the pharmacy.

EXPLAIN WHY

Explain why these statements are true or important. Check your answers in the text. Discuss any questions you may have with your instructor.

1. If an online claim is rejected, why is it important to review the information that was originally entered before calling?

2. Why is it important to know the benefits of various third-party programs?

IN THE WORKPLACE ACTIVITIES

Simulate actions needed to verify and accurately input third-party coverage for a prescription. Answer the questions below for each of the following insurance cards.

1. What is the cardholder ID number?
2. What is the group number?
3. What is the effective date?
4. What number should a pharmacy technician call with questions about a claim?

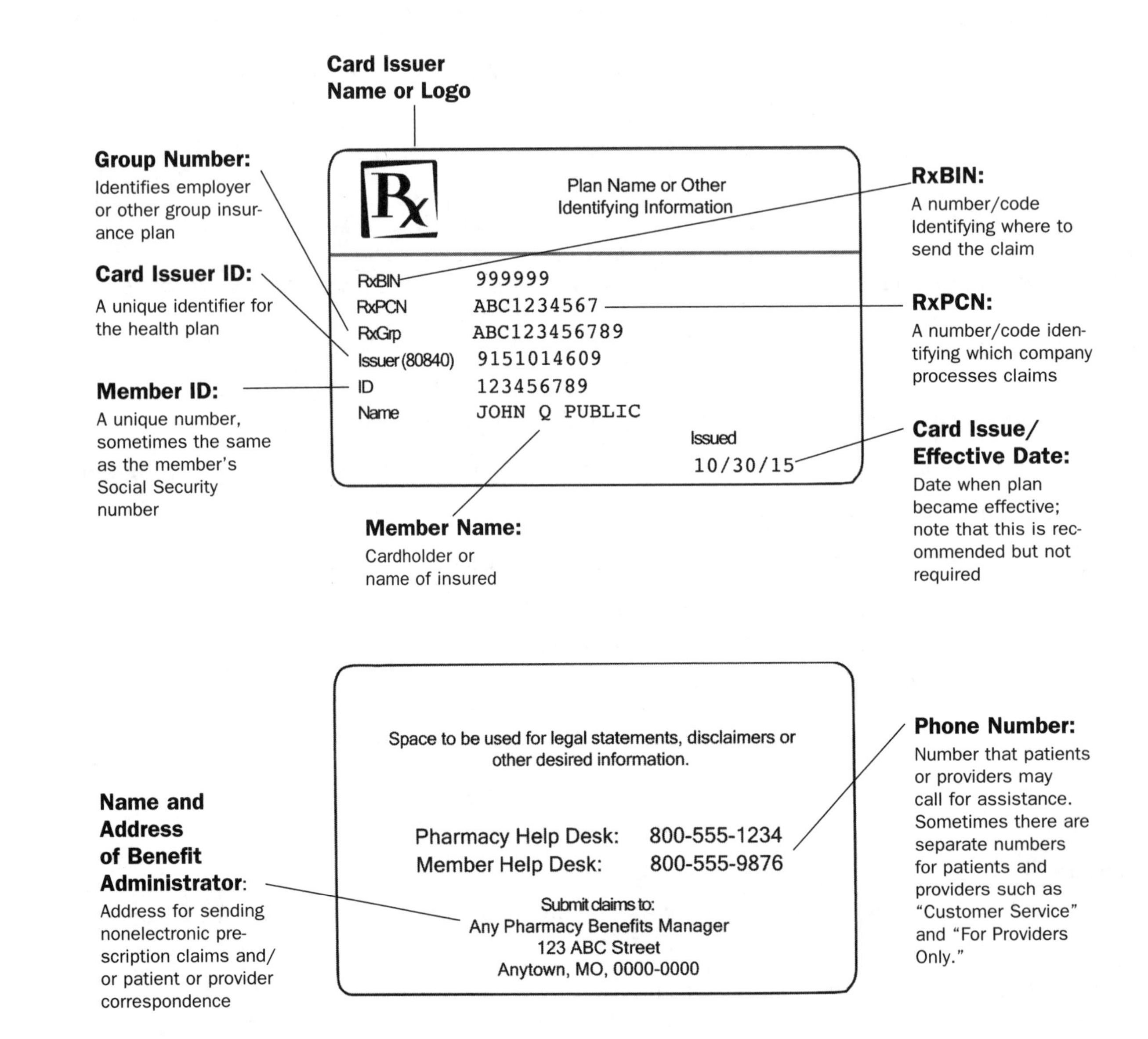

Images courtesy of the National Council for Prescription Drug Programs (NCPDP). Used by permission.

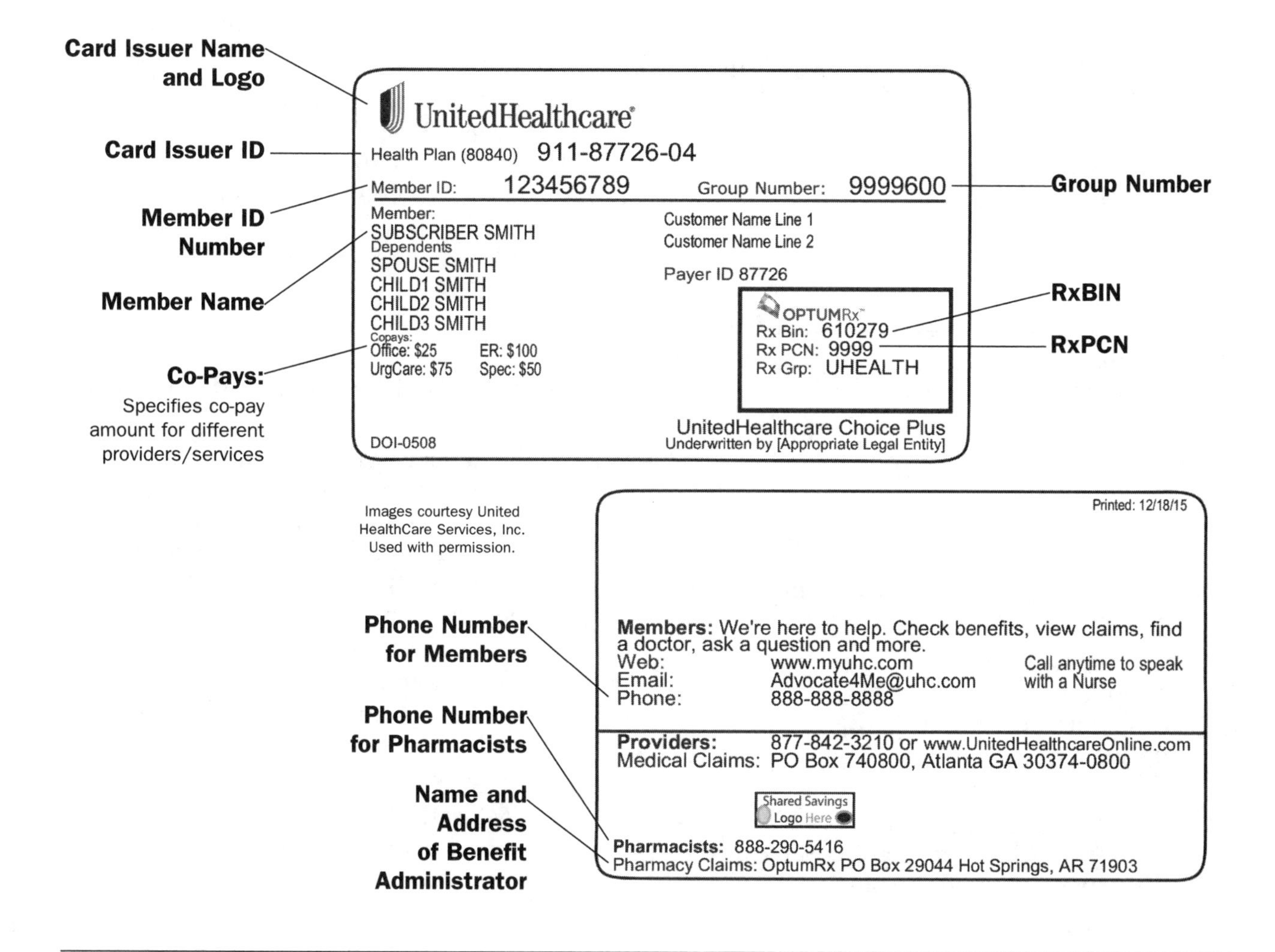

Images courtesy United HealthCare Services, Inc. Used with permission.

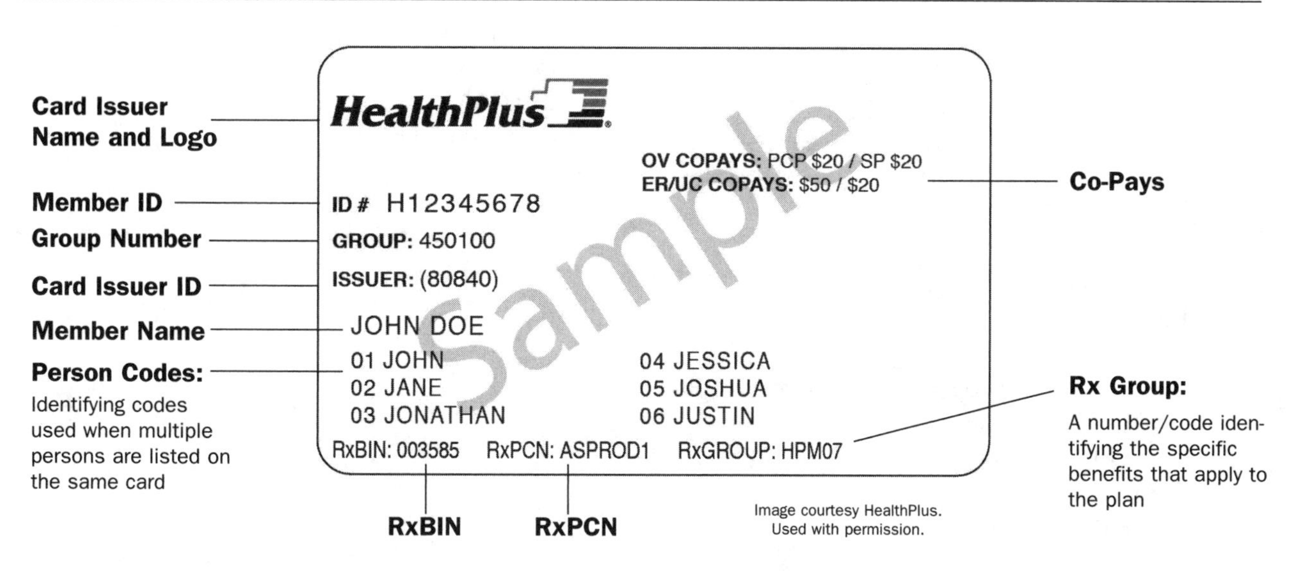

Image courtesy HealthPlus. Used with permission.

CHOOSE THE BEST ANSWER

Answers are at the end of the book.

1. The resolution of prescription coverage for a prescription through the communication of the pharmacy computer with the third-party computer is called
   a. PBM.
   b. online adjudication.
   c. MAC.
   d. UCR.

2. The maximum allowable cost (MAC) is usually _____________ the usual and customary (U&C) price.
   a. less than
   b. greater than
   c. equal to

3. The _____________________ is the maximum price per unit an insurer will pay for a product.
   a. co-insurance
   b. co-pay
   c. maximum allowable cost
   d. deductible

4. A(An) ____________ is a network of providers for which costs are covered inside the network but not outside.
   a. HMO
   b. POS
   c. PPO
   d. MAC

5. _______________ is a program for people over age 65 or with certain disabilities.
   a. Medicaid
   b. ADC
   c. Workers' compensation
   d. Medicare

6. The Medicare program that covers inpatient hospital expenses is
   a. Medicare Part A.
   b. Medicare Part B.
   c. Medicare Part C.
   d. Medicare Part D.

7. Procedures for billing compounded prescriptions
   a. should always be referred to the pharmacist.
   b. are not available.
   c. do not apply in community pharmacy practice.
   d. are variable, depending on the insurer or PBM.

8. When a technician receives a rejected claim "invalid birth date," this probably means
   a. the patient has Medicaid.
   b. the patient does not have coverage.
   c. the birth date submitted by the pharmacy does not match the birth date in the insurer's computer.
   d. the patient has Medicare.

9. The form used by health-care providers to apply for a National Provider Identifier (NPI) is
   a. CMS-1500.
   b. CMS-10114.
   c. a universal claim form.
   d. CPT 0116T.

10. An online platform for MTM.
    a. OutcomesMTM
    b. CHAMPUS
    c. CHAMPVA
    d. TRICARE

STUDY NOTES

Use this area to write important points you'd like to remember.

— 16 —

COMMUNITY PHARMACY

KEY CONCEPTS

Test your knowledge by covering the information in the right-hand column.

community pharmacy	Pharmacies that provides prescription services to the public and sell over-the-counter medications and health and beauty products.
close interaction with patients	A key characteristic of community pharmacy that allows pharmacy technicians to constantly interact with patients.
interpersonal skills	Skills involving relationships between people.
independent pharmacies	Individually owned local pharmacies.
chain pharmacies	Regional or national pharmacy chains such as CVS, Walgreens, and others.
mass merchandiser pharmacies	Regional or national stores such as Walmart, Kmart, Target, and others that have pharmacy departments.
food store pharmacies	Regional or national food store chains such as Kroger, Giant Eagle, and others that have pharmacy departments.
disease state management programs	One-on-one pharmacist-patient consultation sessions to help manage chronic diseases.
MTM	Services that are similar to disease state management, but are tailored to the provisions of Medicare part D. Five core elements of MTM include: medication therapy review, personal medication record, medication related action plan, intervention and/or referral, and documentation and follow up.
walk-in clinics	Clinics in pharmacies that are usually staffed by nurse practitioners and provide treatment for a limited number of common conditions.
pharmacist immunization programs	Depending on state law, pharmacists may administer routine vaccinations.

good customer service	Presenting yourself to customers in a calm, courteous, and professional manner, along with listening to and understanding customer requests for service and accurately fulfilling requests.
federal regulations	Laws such as OBRA, HIPAA, Medicare Prescription Drug Improvement and Modernization Act, CMEA, and the Red Flag Rule that are important for pharmacy technicians to know.
state regulations	Community pharmacies are most closely regulated at the state level.
transaction windows	Counter areas designated for taking prescriptions and for dispensing them to patients.
storage	Adequate shelving, cabinets, drawers, and refrigerator(s) for storing medications.
prescription counter	Counter area designated for preparing noncompounded medications.
compounding area	Counter area, usually near a sink, for preparing medications that require mixing.
prescription bins or shelves	Storage areas for completed prescriptions.
pharmacist's judgement	Some questions and calls require the pharmacist's judgment and these should be directed immediately to the pharmacist.
prescription in-take	The drop-off area where patients bring prescriptions that need to be filled.
patient profile	Information about the patient that is entered into the computer.
online billing	Today's pharmacy systems generally "fill and bill" at the same time.
refills	When processing a refill prescription, it is necessary to check that there are refills available. In the case of a patient requesting an early refill of a controlled substance, involve the pharmacist right away.
partial fills	Dispensing a lesser quantity of a medication when the pharmacy does not have enough in stock to fill a prescription.
transfers	Filling a prescription at another pharmacy other than where it was previously filled. Transfers are generally filled when the transferring pharmacist calls a pharmacist at the pharmacy where the prescription was originally dispensed to get all of the necessary information about the prescription and transfer the remaining refills.
filing	Hard copies of prescriptions are filed by prescription number. Faxed, phoned, and electronic prescriptions must be printed or transcribed onto a paper hard copy so they can be filed.

KEY CONCEPTS (CONT'D)

Test your knowledge by covering the information in the right-hand column.

scanning a hard copy prescription	For accuracy and improved record keeping, many pharmacies also scan the prescription into the pharmacy dispensing system.
vial	Container for dispensing tablets or capsules.
amber bottle	Container for dispensing liquid medications.
safety caps	All dispensed prescription vials and bottles must have a safety cap or child resistant cap, unless the patient requests a non-child resistant cap.
counting tray	A tray designed for counting pills from a stock bottle into a prescription vial.
automated filling and dispensing machines	Machines that automatically fill and label pill bottles with correct quantities of ordered drugs.
reconstitution	Some medications are shipped as powders but must be mixed with distilled water so they are dispensed as liquids.
Fillmaster	A device that dispenses the exact amount of distilled water for reconstitution.
auxiliary labels	Labels regarding specific warnings, foods, or medications to avoid, potential side effects, and other cautionary statements.
final check by the pharmacist	The final step of the prescription preparation process is the final check of the product and all paperwork by the pharmacist.
signature log	Customer signatures in log are required for Medicaid and most third-party insurers, as well as Schedule V controlled substances, poisons, and certain other prescriptions (depending on state laws).
separation and removal of trash	In accordance with HIPAA, pharmacies must separate trash that contains protected health information (PHI). Any paperwork containing PHI must be shredded onsite, or by a contracted vendor.
OTC products	Drug products that do not require a prescription, but are not without risks. Therefore, the technician should not recommend them to pharmacy customers.
markup	The amount of the retailer's sale price minus their purchase price.
shelf stickers	Stickers for OTC drugs and other products that can be scanned for inventory identification.
unit price	The price of a single unit of a product, such as for 1 ounce of a liquid cold remedy.

RETAIL MATH

Technicians in community pharmacies must know how to perform common retail calculations. Besides simple addition and subtraction, the most important calculations involve using percentages, especially in doing markups or discounts. A markup is the amount of the retailer's selling price minus their purchase cost. It is calculated by multiplying the retailer's purchase cost by the markup percentage and adding the amount to the cost.

For example, a 30% markup on an item purchased for $2.30 is $0.69 (Note that 30% equals 0.3, and that $2.30 x 0.3 = $0.69), so the selling price would be $2.99 ($2.30 + $0.69).

Conversely, if you knew a $2.99 sale item was marked up $0.69 and were asked to figure out the percent markup, you would subtract the $0.69 from $2.99 to get the cost of the item ($2.30), and then divide the markup by the cost: $0.69 ÷ $2.30 = 0.3 = 30%.

Discounts involve subtracting a percentage amount from the marked up price of an item. A 30% discount on the $2.99 item is $0.90, so $2.09 would be the discounted price ($2.99 – $0.90). Note that this is different than the cost of the item, because you deducted the percentage from the marked up price.

TRUE/FALSE

Indicate whether the statement is true or false in the blank. Answers are at the end of the book.

_____ 1. Almost two-thirds of all drugs in the United States are dispensed at community pharmacies.

_____ 2. At the counter, listening carefully and making eye contact are good interpersonal techniques.

_____ 3. The pharmacy technician should not recommend OTC products to pharmacy customers.

_____ 4. The dispensing code is only required for controlled substance prescriptions.

_____ 5. HIPAA regulations do not apply to disease state management services provided by pharmacists.

_____ 6. Safety caps are not used for patients who request an easy open cap.

_____ 7. Schedule II controlled substance orders must be checked and signed for by the pharmacist.

_____ 8. When a technician prepares a prescription, it is always checked by the pharmacist before dispensing to the patient.

_____ 9. Signature logs serve as proof to third-party payers that the prescription was dispensed to the patient.

_____ 10. Net profit is the difference between the selling price and the acquisition cost.

_____ 11. Canes, walkers, and wheelchairs are examples of DME.

FILL IN THE KEY TERM

Answers are at the end of the book.

auxiliary labels	**partial fills**	**shelf stickers**
CMEA	**patient profile**	**transaction windows**
disease state management programs	**pharmacist's judgement**	**unit price**
interpersonal skills	**red flag rule**	**walk-in clinics**
markup	**safety caps**	
	signature log	

1. ____________________ : A set of provisions created by the Federal Trade Commission to help prevent identity theft from patient billing accounts that are maintained by medical and financial institutions.
2. ____________________ : Skills involving relationships between people.
3. ____________________ : Clinics in some pharmacies that provide treatment for a limited number of common conditions.
4. ____________________ : Counter areas designated for taking prescriptions and for delivering them to patients.
5. ____________________ : A smaller quantity of medication dispensed when the pharmacy does not have enough in stock to fill a prescription.
6. ____________________ : Some questions and calls require the pharmacist's judgement.
7. ____________________ : Patient information that is entered into the computer.
8. ____________________ : A book or electronic device to provide verification that a prescription was picked up.
9. ____________________ : The amount of the retailer's sales price minus their purchase price.
10. ____________________ : Child resistant caps required of all dispensed prescription vials.
11. ____________________ : Federal legislation enacted to regulate over-the-counter sales of ephedrine and pseudoephedrine.
12. ____________________ : Stickers with bar codes that can be scanned for inventory identification.
13. ____________________ : Provide one-on-one pharmacist-patient consultation to help manage chronic diseases and conditions.
14. ____________________ : Labels regarding specific warnings and usage information.
15. ____________________ : For example, the price for 1 ounce of a liquid cold remedy.

EXPLAIN WHY

Explain why these statements are true or important. Check your answers in the text. Discuss any questions you may have with your instructor.

1. Why is the health of customers a factor in community pharmacy?

2. Why are good interpersonal skills important in community pharmacy?

3. Why is it important to look a patient in the eye and restate what they have said?

4. Why does the pharmacist check technician-filled prescriptions before dispensing to patients?

5. Why shouldn't technicians recommend OTC products?

IN THE WORKPLACE ACTIVITIES

1. For the drugs in the prescriptions on pages 70–75, what product should be selected if the patient's prescription card only covers generic?
2. In accordance with your state tax laws, determine which of the following products are taxable: Tylenol®, B-D alcohol swabs, One Touch test strips, Bayer® aspirin, Scope mouthwash, Crest toothpaste.
3. Role-play cash register procedures and actually make change for purchases.
   a. Cost of prescription is $65.28 and patient pays with $80
   b. Cost of prescription is $9.97 and patient pays with $20
   c. Cost of prescription is $15.50 and patient pays with $20
   d. Cost of prescription is $22.55 and patient pays with $50
4. Role-play with two students obtaining a patient profile for a new patient who is getting prescriptions filled for the first time at your pharmacy. What information is needed? Is there a form you can use to make sure that you get all the necessary information?
5. Review the ISMP website (www.ismp.org) Simulate a medication error and respond appropriately. View the website to report errors to ISMP https://www.ismp.org/errorReporting/reportErrortoISMP.aspx. Discuss when it would be appropriate to report an error to ISMP.
6. Working in pairs, check-in a schedule II order, and record the information on DEA Form 222. at www.deadiversion.usdoj.gov/pubs/manuals/narcotic/appendixb/222a.htm.
7. Work with a partner and create a table with the names of five currently available blood glucose monitors. Identify which test strips should be used with each monitor. Identify which lancet device could be used. Identify which lancets should be used with the lancet device.
8. Work with a partner. Review the document from Outcomes MTM on the technician's role in medication therapy management at http://c.ymcdn.com/sites/www.wsparx.org/resource/resmgr/imported/0830PHARMTECHGUIDE.pdf. Discuss some practical considerations.

IN THE WORKPLACE

Use this tool to practice and check your interpersonal skills at the pharmacy counter.

Pharmacy Technician Skills Checklist
INTERPERSONAL TECHNIQUE: COUNTER

Name: ____________________

	Self-Assessment		Supervisor Review		
Skill or Procedure	Needs to Improve	Meets or Exceeds	Needs to Improve	Meets or Exceeds	Plan of Action
1. Listens carefully.					
2. Makes eye contact.					
3. Repeats what the patient/customer said.					
4. Uses positive language to describe what you can do.					
5. Calls the patient by name.					

IN THE WORKPLACE

Use this tool to practice and check your interpersonal skills on the telephone.

Pharmacy Technician Skills Checklist

INTERPERSONAL TECHNIQUE: TELEPHONE

Name: ______________________________

Skill or Procedure	Self-Assessment		Supervisor Review		
	Needs to Improve	Meets or Exceeds	Needs to Improve	Meets or Exceeds	Plan of Action
1. Uses a pleasant voice and courteous manner.					
2. States the name of the pharmacy and self.					
3. Follows the standard procedure indicated for pharmacy.					
4. Refers all calls that require a pharmacist's judgment to the pharmacist.					

IN THE WORKPLACE

Use this tool to practice and check your basic computer entry skills.

Pharmacy Technician Skills Checklist
BASIC COMPUTER ENTRY SKILLS

Name: ____________________

Skill or Procedure	Self-Assessment		Supervisor Review		
	Needs to Improve	Meets or Exceeds	Needs to Improve	Meets or Exceeds	Plan of Action
1. Properly interprets medical abbreviations.					
2. Understands dosage forms.					
3. Understands prescription number assignments (controlled, non-controlled).					
4. Properly enters a new patient into computer.					
5. Properly adds a new patient to an existing family file.					
6. Properly uses SIG codes.					
7. Properly adds medication allergy codes.					
8. Properly processes refill prescriptions using computer.					
9. Understands procedures for DUR screens.					
10. Properly enters third-party billing (insurance) information.					
11. Properly adds a new doctor file (if doctor is not already in the database).					
12. Chooses correct drug (matches NDC).					
13. Uses appropriate DAW codes.					

IN THE WORKPLACE

Use this tool to practice and check your general computer entry skills.

Pharmacy Technician Skills Checklist
GENERAL COMPUTER ENTRY SKILLS

Name: ______________________

Skill or Procedure	Self-Assessment: Needs to Improve	Self-Assessment: Meets or Exceeds	Supervisor Review: Needs to Improve	Supervisor Review: Meets or Exceeds	Supervisor Review: Plan of Action
1. Properly puts new prescriptions on file (without filling).					
2. Enters and uses alternate third-party plans.					
3. Enters compounded prescriptions.					
4. Generates daily reports.					
5. Generates medical expense statements.					
6. Updates inventories.					
7. Updates order quantities.					

CHOOSE THE BEST ANSWER

Answers are at the end of the book.

1. Community pharmacies within stores, like Walmart or Kmart, that are part of regional or national mass merchandise chains are
   a. chain pharmacies.
   b. mass merchandiser pharmacies.
   c. independent pharmacies.
   d. food store pharmacies.

2. ________ requires community pharmacists to provide counseling to Medicaid patients.
   a. OBRA 90
   b. HIPAA
   c. MMA
   d. CMEA

3. MTM is part of
   a. OBRA 90.
   b. HIPAA.
   c. MMA.
   d. CMEA.

4. Medicare Part D relies on _________ to provide medication coverage to eligible patients.
   a. Medicaid
   b. patient assistance programs
   c. Prescription Drug Plans
   d. state welfare programs

5. The Red Flag Rule was enacted to
   a. help prevent identity theft.
   b. limit sales of pseudoephedrine.
   c. create PDPs.
   d. provide counseling to Medicaid patients.

6. The final check of a new prescription is performed by
   a. the pharmacist.
   b. the prescribing physician.
   c. a certified technician.
   d. the senior technician.

7. Scanning a hard-copy prescription is
   a. required by OBRA 90.
   b. required by HIPAA.
   c. for purposes of accuracy and record-keeping.
   d. required by MMA.

8. OTC product recommendations should be made by
   a. certified technicians.
   b. technicians with seniority.
   c. pharmacists.
   d. technicians with more than two years of experience.

9. Required to be dispensed with oral contraceptives:
   a. MedGuides.
   b. PPIs.
   c. Auxiliary labels.
   d. Safety cap.

10. Pharmacy technicians are responsible for all of the following EXCEPT:
    a. reordering stock.
    b. recommending OTC products to patients.
    c. keeping the pharmacy neat and clean.
    d. product stock duties.

11. Separation of trash in pharmacies is in accordance with
    a. HIPAA.
    b. CMEA.
    c. OBRA.
    d. PDP.

STUDY NOTES

Use this area to write important points you'd like to remember.

– 17 –

HOSPITAL PHARMACY

KEY CONCEPTS

Test your knowledge by covering the information in the right-hand column.

pharmacist supervision Pharmacy technicians in the hospital work under the direct supervision of a pharmacist or supervising technician. Only a pharmacist may verify orders in the computer system and check medications being sent to the nursing floors.

patient care units Patient rooms are divided into groups called nursing units or patient care units, with patients having similar problems often located on the same unit.

nurse's station The work station for medical personnel on a nursing unit is called the nurse's station. Various items required for care of patients are stored there, including patient medications.

ancillary areas Areas such as the emergency room that also use medications and are serviced by the pharmacy department.

central supply An area of the hospital that may carry supplies not provided by the pharmacy such as lotion and mouthwash.

central pharmacy The main inpatient pharmacy in a hospital that has pharmacy satellites.

front counter Area near the entrance of the main pharmacy where pharmacy technicians help other health-care professionals, answer phone calls, fill first doses of oral medications, and perform other duties as needed.

satellite pharmacy A branch of the inpatient pharmacy responsible for preparing, dispensing, and monitoring medication for specific patient areas.

delivery technician A delivery technician is responsible for transporting medications and other pharmacy supplies from the pharmacy to nursing units, ancillary areas of the hospital, and/or outpatient clinics.

outpatient pharmacy A pharmacy attached to a hospital servicing patients who have left the hospital or who are visiting doctors in a hospital outpatient clinic.

order processing	Entering written medication orders in the computer system.
monitoring drug therapy	Retrieving drug levels, lab values, or other patient specific information from patient charts or electronic records to assist a pharmacist.
investigational drug service	A specialized pharmacy subsection that deals solely with clinical drug trials. These drug studies require a great deal of paperwork and special documentation of all doses of study medication taken by patients. Technicians frequently assist the pharmacist with this documentation and in preparing individual patient medication supplies.
inventory control	Maintaining the stock of drugs and supplies in the pharmacy.
cart fill	In hospitals that have manual cart fill, medication carts contain a 24-hour supply of medications in a patient cassette or drawer.
automation	Pharmacy systems, including robots and automated dispensing cabinets.
narcotics/controlled substances	Pharmacy technicians can help coordinate narcotic drug distribution, including reviewing reports and confirming compliance with state and federal controlled substance laws.
IV/clean room	Areas designated for the preparation of sterile products.
quality assurance	Involves inspecting nursing units and other areas of the hospital that store medications to make sure medications are stored and handled in compliance with hospital policy.
chemotherapy	Additional training is important on handling and preparing hazardous materials including chemotherapy drugs.
pharmacy technician supervisor	Part of the management team responsible for training technicians, creating work schedules, and completing annual evaluations.
formulary	A list of drugs stocked at the hospital that have been selected based on therapeutic factors as well as cost.
closed formulary	A closed formulary requires physicians to order medications from the formulary list.
non-formulary	Drugs not on the formulary list.
therapeutic interchange	A policy approved by the hospital pharmacy & therapeutics committee that allows the pharmacist to change a medication order to a therapeutically equivalent formulary medication.
pneumatic tube	A system that shuttles objects through a tube using compressed air as the force.
electronic medical record	A computerized patient medical record, also known as electronic health record.

KEY CONCEPTS (CONT'D)

Test your knowledge by covering the information in the right-hand column.

medication order form In the hospital, all drugs ordered for a patient are written on a medication order form and not a prescription blank as in a community pharmacy. Physicians write medication orders for hospital patients, though both nurses and pharmacists may also write orders if they are directly instructed to do so by a doctor. In addition, physician's assistants and nurse practitioners may sometimes write orders, depending upon the institution.

medication administration record Nurses record and track medication orders on a patient specific form called the medication administration record (MAR).

computerized physician order entry (CPOE) A computer system that allows the physician to enter the medication order directly into the hospital computer system.

standing order A standard medication order for patients to receive medication at scheduled intervals.

PRN order Orders for medications that are administered only on an as-needed basis.

STAT order An order for medication to be administered immediately.

unit dose A package containing the amount of drug for one dose.

prepacking Technicians often "prepack" medications that have been supplied in bulk into unit doses. Machines that automate this process are generally used for prepacking oral solid medications.

unit dose labels Unit dose labels contain bar codes for identification and control. Items are scanned into the dispensing and inventory system at various stages up to dispensing. This reduces the chances of medication errors and improves documentation and inventory control.

IV admixtures A large portion of the medication used in the hospital is administered intravenously. Pharmacy technicians prepare IV admixtures, including small and large volume parenterals, parenteral nutrition therapy, and chemotherapy.

infection control committee Hospital committee in charge of the surveillance, prevention, and control of infection within the hospital.

TJC The Joint Commission on Accreditation of Healthcare Organizations, the accreditation agency for health-care organizations. Organizations undergo a TJC survey every three years.

code cart Locked cart filled with emergency medications. All patient care areas are required to have code carts.

proper waste disposal In order to keep workers, the community, and the environment safe, medications and medical supply waste must be handled properly.

sharps container Needles or other items that may cut or puncture the skin should always be thrown away in designated sharps containers.

emergency codes Color-based emergency code systems that are used at many hospitals to alert staff to various emergency and safety situations.

policy and procedures manual A manual containing information about every aspect of the job from dress code to disciplinary actions and step-by-step directions on how to perform various tasks that will be required of technicians. All departments within the hospital are required by regulating agencies to maintain this.

---

unit dose medications

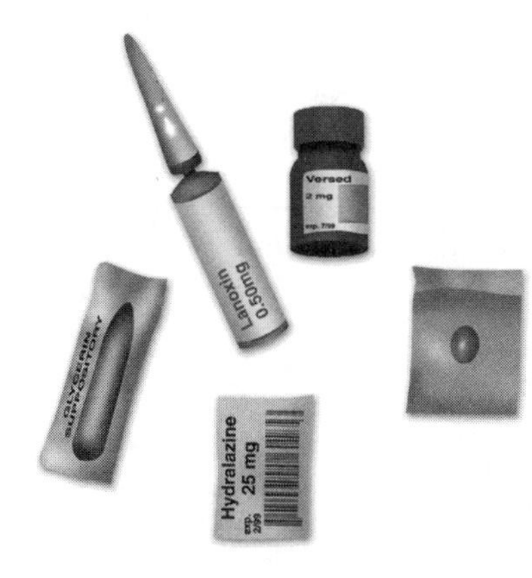

patient trays

MEDICATION ORDERS

Medication order forms are an all-purpose communication tool used by the various members of the health-care team. Orders for various procedures, laboratory tests, and X-rays may be written on the form in addition to medication orders. Several medication orders may be written on one medication order form unlike pharmacy prescription blanks seen in the retail setting.

MEMORIAL HOSPITAL
5/12/2016 2200
Admit patient to 6th floor
dx: Pneumonia, dehydration
All: PCN - rash
Order CBC, chem-7, blood culture STAT
NS @ 125mL/hr IV Dr. Johnson x2222
Smith, John 099999999 12/4/1950

MEMORIAL HOSPITAL
5/13/2016 0300
tylenol 650mg po q4h prn for temp >38°C
Dr. Johnson x2222

There are several different types of orders that can be written. One is a standard medication order for patients to receive a certain drug at scheduled intervals throughout the day, sometimes called a standing order. Orders for medications that are administered only on an as-needed basis are called PRN medication orders. A third type of order is for a medication that is needed right away and these are referred to as STAT orders.

MEDICATION ADMINISTRATION RECORD

On this form every medication ordered for a patient is written down as well as the time it is administered and the person who gave the dose. These forms may be handwritten by the nursing staff or generated by the pharmacy computer system. The MAR is an important document in tracking the care of the patient because it gives a 24-hour picture of a patient's medication use. The accuracy of this document is crucial.

COMMUNITY HOSPITAL
Medication Administration Record

Room/Bed: 675-01
Patient: SMITH, JOHN
Account #: 099999999
Sex: M
Age: 65 Y
Doctor: JOHNSON, P.
From 0730 on 02/01/16 to 0700 on 02/02/16
Diagnosis: PNEUMONIA; DEHYDRATION
Height: 5'11" weight: 75KG
Verified By: Susie Smith, RN
Allergies: PENICILLIN-->RASH

	0730–1530	1600–2300	2330–0700
0.9% SODIUM CHLORIDE 1 LITER BAG DOSE 125 ML/HR IV ORDER #2	800 JD	1600 SS	2400
MULTIVITAMIN TABLET DOSE: 1 TABLET PO DAILY ORDER #4	1000 Given 9AM JD		
CLARITHROMYCIN 500 MG TABLET DOSE: 500 MG PO Q 12 HRS ORDER #5	1000 JD	2200 SS	
ACETAMINOPHEN 325 MG TABLET DOSE: 650 MG PO Q 4-6 HRS PRN FOR TEMP>38° C ORDER # 17	1200 JD		

Init / Signature
SS / Susie Smith, RN
JD / Jane Doe, RN

Init / Signature

FILL IN THE KEY TERM

Answers are at the end of the book.

central pharmacy
clean rooms
code carts
CPOE
electronic medical record
inpatient pharmacy
medication administration record (MAR)
outpatient pharmacy
par
pharmacy satellite
policy and procedures manual
PRN order
reconstitute
standard precautions
standing order
STAT order
unit dose

1. ____________________ : Addition of water or other diluent to a powdered drug form to make a solution or suspension.
2. ____________________ : A system in which the prescriber enters orders directly into the computer system.
3. ____________________ : A computerized patient medical record.
4. ____________________ : A package containing the amount of a drug required for one dose.
5. ____________________ : A standard medication order for patients to receive medication at scheduled intervals.
6. ____________________ : An order for medication to be administered only on an as-needed basis.
7. ____________________ : An order for medication to be administered immediately.
8. ____________________ : A form that tracks the medications administered to a patient.
9. ____________________ : A locked cart of medications designed for emergency use only.
10. ____________________ : The main inpatient pharmacy in a hospital that has satellite pharmacies.
11. ____________________ : A pharmacy located in a hospital that serves only those patients in the hospital and its ancillary areas.
12. ____________________ : A branch of the inpatient pharmacy.
13. ____________________ : The amount of drug product that should be kept on the pharmacy shelf.
14. ____________________ : Areas designed for the preparation of sterile products.
15. ____________________ : A pharmacy attached to a hospital servicing patients who have left the hospital or who are visiting doctors in a hospital outpatient clinic.

FILL IN THE KEY TERM (CONT'D)

Answers are at the end of the book.

16. ____________________ : Documentation of required policies, procedures, and disciplinary actions in a hospital.

17. ____________________ : The practice of avoiding direct contact with blood, mucous membranes, body fluids, and non-intact skin by the use of personal protective equipment.

TRUE/FALSE

Indicate whether the statement is true or false in the blank. Answers are at the end of the book.

_____ 1. Technicians working the front counter do not handle phone calls.

_____ 2. Medication carts in hospitals contain a 72–hour supply of medications.

_____ 3. Technicians often prepare IV admixtures in hospitals.

_____ 4. The pharmacy technician supervisor is often responsible for training technicians.

_____ 5. Lot numbers and expiration dates are not needed in bulk compounding logs.

_____ 6. All departments within a hospital are required to maintain a policy and procedure manual.

_____ 7. The pharmacy inventory staff is often responsible for removal of drug recalls from the pharmacy inventory.

EXPLAIN WHY

Explain why these statements are true or important. Check your answers in the text. Discuss any questions you may have with your instructor.

1. Why can several medication orders be written on a single medication order form?

2. Why is it important for technicians to be familiar with the policy and procedures manual for their department?

IN THE WORKPLACE

Some questions to ask when entering or checking a medication order.

Several questions should always be asked when entering or checking a medication order.

Question:	Example situation	
	Problem	Solution*
✓ Is the patient allergic to the ordered medication or a component of the ordered medication?	A medication order is placed for trimethoprim/sulfamoxazole (Bactrim®), but the patient has a "sulfa" allergy with the description of the allergy stating "hives".	The order should not be processed and a new medication order for a different antibiotic is required.
✓ Is the drug appropriate?	A patient is ordered a potassium IVPB. The most recent potassium lab level = 5 mEq/mL (normal potassium = 3.5 to 5.5 mEq/mL). The medication is not indicated for this patient.	The physician should be notified that the medication is not indicated based on the current lab level.
✓ Is the dose and frequency appropriate?	An order for a patient reads, "Digoxin 0.5 mg po q 6 H" The normal dosing for digoxin is 0.125 – 0.25 mg daily.	The ordered dose would be toxic. The order should be changed.
✓ Has the drug dose/frequency been adjusted for the patient's age, weight, disease state, kidney or liver function?	A 79 year old patient with major renal dysfunction has a new medication order for Fluconazole 400 mg IV daily. Fluconazole is eliminated renally and should be adjusted for renal dysfunction.	The patient's renal function should be calculated and the dose adjusted appropriately.
✓ Is the drug formulation and route most appropriate for the patient?	A patient has an order for an IV medication but the patient is tolerating oral feeds.	Discuss with the physician and change the drug to the oral formulation.
✓ Are there any drug interactions that require changes in the order or increased monitoring that the physician should be made aware?	A new order for amiodarone is received for a patient who is also on warfarin. There is an interaction between amiodarone and warfarin which increases the risk for the patient to bleed.	The physician should be made aware of the interaction and recommended to monitor the patient more closely.
✓ Does the new order therapeutically duplicate another drug the patient is already taking?	A medication order for pantoprazole is ordered for a patient with currently active order for omeprazole. Both medications are proton pump inhibitors and work in the same way.	There would be no additive effect to using both medications. The order needs to be clarified with the physician and one of the orders discontinued.
✓ Is the drug on the hospital formulary? If not, can the medication be therapeutically interchanged?	Famotidine 20mg BID is ordered but the formulary H_2-blocker is ranitidine.	A therapeutic substitution is made to ranitidine 150 mg BID per hospital protocol.
✓ Is the medication "restricted" and need approval from a specific service in the hospital?	Vancomycin 1 g IV q 12 H is ordered for a patient. Vancomycin is restricted by the infectious disease service and it has not yet been approved for this patient.	The infectious disease service should be notified and approval received before processing the order.
✓ Should the medication be limited to a certain number of days due to hospital policy or an automatic stop order?	An order is written for ketoralac 30 mg IV q 6H x 10 days. Hospital policy and manufacturer recommendations limit the length of therapy to 5 days.	The order is changed to 5 days of therapy.

*Unless there is a hospital policy in place to the contrary, all changes in medication orders need to be authorized by the physician before they are changed.

IN THE WORKPLACE

A sample extemporaneous compounding worksheet.

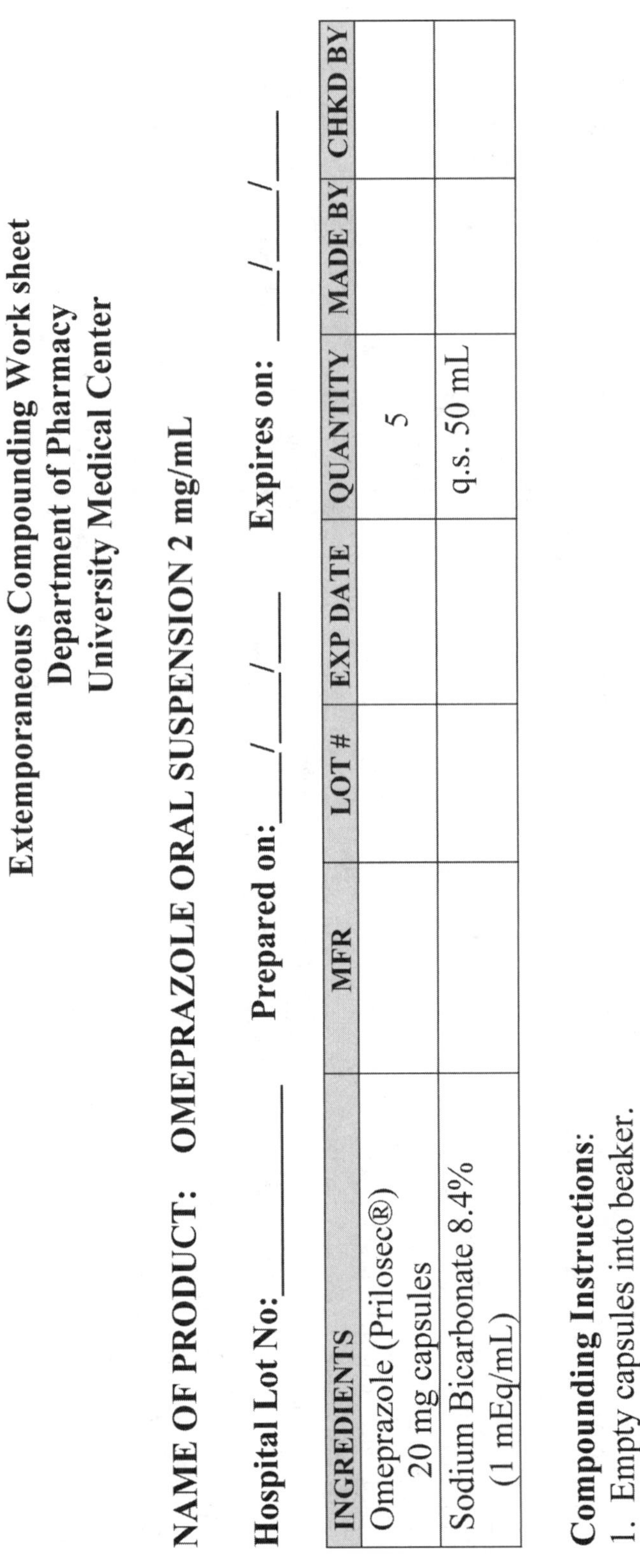

Extemporaneous Compounding Work sheet
Department of Pharmacy
University Medical Center

NAME OF PRODUCT: OMEPRAZOLE ORAL SUSPENSION 2 mg/mL

Hospital Lot No:__________ **Prepared on:**___/___/___ **Expires on:** ___/___/___

INGREDIENTS	MFR	LOT #	EXP DATE	QUANTITY	MADE BY	CHKD BY
Omeprazole (Prilosec®) 20 mg capsules				5		
Sodium Bicarbonate 8.4% (1 mEq/mL)				q.s. 50 mL		

Compounding Instructions:
1. Empty capsules into beaker.
2. Add sodium bicarbonate solution.
3. Gently stir (about 15 minutes) until a white suspension is formed.
4. Transfer to amber-colored bottle.

Stability:
45 days at 4°C

Auxiliary Labels required:
SHAKE WELL
REFRIGERATE
PROTECT FROM LIGHT

References:
Ann Pharmacother, 2000;34:600-605.
Am J Health Syst Pharm, 1999;56(suppl 4): S18-21

IN THE WORKPLACE

Use this tool to practice and check your skill in preparing a parenteral.

Pharmacy Technician Skills Checklist
PREPARING A PARENTERAL

Name: ______________________________

Skill or Procedure	Self-Assessment		Supervisor Review		
	Needs to Improve	Meets or Exceeds	Needs to Improve	Meets or Exceeds	Plan of Action
1. The technician calculates the amount of drug needed for the IVPB (intravenous piggyback).					
2. The required equipment is gathered: medication vial, base solution bag, syringes, needles, and alcohol swabs.					
3. Prior to preparation, hands should be washed, gloves and gown worn, and laminar flow hood cleaned.					
4. The IVPB is prepared using proper aseptic technique to ensure sterility.					
5. A double check should always be completed: check for correct drug, dose, concentration, volume, base solution, and all calculations.					
6. The technician should write his/her initials on the label and add an expiration date and time prior to placing the label on the IVPB.					
7. Then the vial and syringes used in preparation are placed next to the IVPB for the pharmacist to check.					
8. All supplies used for product preparation should be disposed of in the proper bins as required by the hospital pharmacy.					

IN THE WORKPLACE

A sample orientation checklist.

UNIVERSITY MEDICAL CENTER
Orientation Checklist

NAME: ____________________ POSITION: ________________ HIRE DATE: ____________

Pharmacy ORIENTATION SKILL CHECKLIST (outline of competency requirements)	EVALUATION			
	Reviewed (date/initial)	Verbalizes Understanding (date/initial)	Demonstrates Skill (date/initial)	COMMENTS
INTRODUCTION AND TOUR Introduce to all personnel				
Tour pharmacy department - Central pharmacy and pharmacy satellites				
Tour main pathways from department to different areas in hospital				
GENERAL INFORMATION Mailbox				
Phone etiquette/personal calls				
Regular and Holiday schedule				
Vacation and Sick time off				
Time clock (punching in/out)				
PERSONAL NEEDS Dress code				
Cloak storage/locker				
Lunch and coffee breaks				
SECURITY AND SAFETY Exposure control				
Proper waste disposal				
Material Safety Data Sheets (MSDS)				
Mass Casualty Disaster Plan				
COMPUTER ENVIRONMENT / ORDER PROCESSING Computer system (CPOE, Entry and Retrieval, Allergy, Patient Profiles, etc.)				
Order Verification (R.Ph. only)				
Printing labels				
Filling orders				
Managing missing doses				
Delivery schedule/tube system				
Restocking automated dispensing cabinets				
Down-time procedures				
POLICY AND PROCEDURE MANUAL Receipt of manual				
EMPLOYEE HANDBOOK Receipt of handbook				
JOB DESCRIPTION **PERFORMANCE EVALUATION**				

Pharmacy ORIENTATION SKILL CHECKLIST (outline of competency requirements)	EVALUATION			
	Reviewed (date/initial)	Verbalizes Understanding (date/initial)	Demonstrates Skill (date/initial)	COMMENTS
IV SOLUTIONS Clean room usage and monitoring				
IV solution preparation procedures				
Aseptic technique				
IV Scheduling – 12-hour batching				
IV area restocking routine				
Total Parenteral Nutrition solutions				
Chemotherapy preparation				
Checking IV solutions (RPh only)				
IV admixture infection control consideration				
MISCELLANEOUS Narcotic replenishment				
Narcotic delivery/Narcotics in IV Rom				
Inventory requests				
Servicing ancillary areas				
Borrowing/lending				
Outpatient pharmacy services – Hours of operation				
Crash cart trays/crash cart re-issue				
Investigational drugs				
Repackaging request				
Ethyl alcohol requisitions				
Inventory check/order process				
Requisitioning from General Stores and Materials Management				
Restricted drug programs/Approvals				
On-call coordinator, clinical staff, Pharmacist-In-Charge				
Non-formulary drugs				
Pager system				
Unit-dose cassette process and exchange (Adults/Peds)				

37-0111.1

Most hospital pharmacies utilize a "Pharmacy Technician Orientation Checklist." Once each skill has been demonstrated, the checklist is initialed by the technician and the supervisor. Upon completion it is placed in the pharmacy technician's personnel file.

IN THE WORKPLACE ACTIVITIES

1. Work with a partner and practice special procedures regarding investigational drugs in a simulated setting.
2. Work with a partner and practice applying special handling procedures for drugs with mandated Risk Evaluation and Mitigation Strategies (REMS). See http://www.fda.gov/Drugs/DrugSafety/PostmarketDrugSafetyInformationforPatientsandProviders/ucm111085.htm and http://www.accessdata.fda.gov/scripts/cder/rems/index.cfm.
3. Work with a partner and review the ASHP document and investigational drug handling and procedures. See http://www.ashp.org/DocLibrary/BestPractices/ResearchGdlClinical.aspx.

CHOOSE THE BEST ANSWER

Answers are at the end of the book.

1. An area of the hospital that may carry supplies not provided by the pharmacy:
   a. Central pharmacy.
   b. Central Supply.
   c. ASC.
   d. PAR.

2. A committee that assures any hospital research complies with federal, hospital, and ethical standards:
   a. PCT.
   b. IRB.
   c. QA.
   d. PACU.

3. The main inpatient pharmacy in the hospital:
   a. Pharmacy satellite.
   b. Clean room.
   c. Central Supply.
   d Central pharmacy.

4. Includes preparing and delivering medications for drug studies:
   a. Investigational drug service.
   b. Unit dose.
   c. Cart fill.
   d. Inventory control.

5. A pharmacy attached to a hospital that serves patients who have left the hospital or who are visiting doctors in a hospital outpatient clinic is a(an)
   a. inpatient pharmacy.
   b. pharmacy satellite.
   c. central pharmacy.
   d. outpatient pharmacy.

6. Pharmacy technicians working in this satellite require special training and chemotherapy certification from the pharmacy department.
   a. Oncology
   b. Pediatric
   c. OR
   d. Radiology

7. Initial orders when a patient is admitted to a hospital:
   a. Standing orders.
   b. PRN orders.
   c. STAT orders.
   d. Admission orders.

8. The unit dose package type used for ointments and creams is a(an)
   a. plastic blister.
   b. ampule.
   c. tube.
   d. vial.

9. Orders for medications that are needed right away are called
   a. PRN orders.
   b. parenteral.
   c. STAT orders.
   d. standing orders.

10. A locked cart of medications designed for emergency use is called a
   a. PCU.
   b. code cart.
   c. satellite.
   d. IP.

11. Rooms designed for the preparation of sterile products are called
   a. satellites.
   b. clean rooms.
   c. CPs.
   d. PCUs.

12. A medication safety event that had potential to cause harm but did not reach the patient:
   a. Sentinel event.
   b. Close call.
   c. Incident.
   d. Medical error.

STUDY NOTES

Use this area to write important points you'd like to remember.

— 18 —

OTHER ENVIRONMENTS

KEY CONCEPTS

Test your knowledge by covering the information in the right-hand column.

mail order pharmacy Delivery of prescriptions by mail (primarily for maintenance therapy). Mail order pharmacies are generally large-scale operations that are highly automated. They use assembly line processing in which each step in the prescription fill process is completed or managed by a person who specializes in that step.

maintenance therapy Therapy for chronic conditions that include depression, gastrointestinal disorders, heart disease, hypertension, and diabetes.

regulation Mail order pharmacies must follow federal and state requirements in processing prescriptions, but are not necessarily licensed in each state to which they send medications.

pharmacist review Pharmacists review mail order prescriptions before and after filling.

online drugstore A type of mail order pharmacy that uses the Internet to advertise and take orders for drugs.

home care Care in the home, generally supervised by a registered nurse who works with a physician, pharmacist, and others to administer a care plan that involves the patient or another caregiver.

home infusion Infusion administered in the home, the fastest growing area of home health care. The primary therapies provided by home infusion services are: antibiotic therapy, parenteral nutrition, pain management, and chemotherapy.

infusion pumps Pumps that control infusion. There are pumps for specific therapies or multiple therapies, as well as ambulatory pumps that can be worn by patients.

patient education In home infusion, the patient or their caregiver is educated about their therapy: how to self-administer, monitor, report problems, and so on.

admixture preparation	The same rules apply to preparing parenteral admixtures in the home infusion setting as in the hospital.
long-term care	Facilities that provide care for people unable to care for themselves because of mental or physical impairment. Because of limited resources, most long-term care facilities will contract out dispensing and clinical pharmacy services.
distributive pharmacist	A long-term care pharmacist responsible for making sure patients receive the correct medicines that were ordered.
consultant pharmacist	A long-term care pharmacist who develops and maintains an individualized pharmaceutical plan for every long-term care resident.
emergency kits	Locked kits containing emergency medications, similar to code carts used in hospitals.
automated dispensing systems	Automated units that dispense medications at the point of use.
nuclear pharmacy	Specially licensed and regulated pharmacies that prepare and dispense radiopharmaceuticals.
compounding community pharmacy	Pharmacies that specialize in compounding services but may also fill prescriptions for commercially available products and/or sell some over-the-counter products; sometimes called hybrid pharmacies.

TRUE/FALSE

Indicate whether the statement is true or false in the blank. Answers are at the end of the book.

_____ 1. Generally drugs cost less in the United States than in Canada.

_____ 2. Mail order pharmacies are often used to handle maintenance medications.

_____ 3. Code carts are never found in long-term care facilities.

_____ 4. Antibiotic therapy is a common home infusion service used in treating AIDS related and other infections.

_____ 5. On the home-care team, the technician works under the supervision of a home-care aide.

_____ 6. Institutional care always provides better quality of life for the patient than home care.

_____ 7. Nuclear pharmacies typically operate 365 days a year.

_____ 8. PET scans use radiopharmaceuticals.

EXPLAIN WHY

Explain why these statements are true or important. Check your answers in the text. Discuss any questions you may have with your Instructor.

1. Why is mail order pharmacy growing so rapidly?
2. Why would maintenance drugs be well suited to mail order delivery?
3. Why is patient education important in home infusion?
4. Why is home infusion growing so rapidly?
5. Why is storage of admixtures an issue in home infusion?
6. Why is hazardous waste an issue in home infusion?

IN THE WORKPLACE ACTIVITIES

1. Prepare a resume for a job as a pharmacy technician in one of the following settings: community pharmacy, hospital pharmacy, long-term care pharmacy, mail order pharmacy, compounding pharmacy.
2. Role-play with three students. One student is the pharmacist, one student is the technician, and one student is the home-care nurse. The home-care nurse calls the pharmacy because the patient has a change in her medication. Think about what steps are needed in order to make the change.

CHOOSE THE BEST ANSWER

Answers are at the end of the book.

1. In the United States, a mail order pharmacy can provide services to
   a. only the states where the company has pharmacies.
   b. only the state of the main campus.
   c. only the state of the main campus and adjacent states.
   d. any state in the United States
2. A medication that is required on a continuing basis for the treatment of a chronic condition is called
   a. PRN medication.
   b. PO medication.
   c. STAT medication.
   d. maintenance medication.

3. Medication counseling for mail order pharmacies is done by
   a. registered nurses.
   b. certified technicians.
   c. nurses.
   d. pharmacists.

4. In some long-term care facilities the medication carts are usually filled with enough medications to last
   a. one week.
   b. one day.
   c. one month.
   d. 24 hours.

5. Medications used from emergency kits in long-term care pharmacies are charged to
   a. the patient.
   b. the pharmacy.
   c. the long-term care facility.
   d. none of the above.

6. The type of infusion therapy that usually involves infusion of narcotics for patients with painful terminal illness or severe chronic pain is called
   a. laxative therapy.
   b. maintenance therapy.
   c. pain management therapy.
   d. infection therapy.

7. Regulation of hazardous waste procedures associated with chemotherapy applies to all of the following EXCEPT:
   a. storage.
   b. disposal.
   c. transportation.
   d. pricing.

8. Which member of the home care team is responsible for educating the patient?
   a. R.Ph.
   b. Physician
   c. Registered nurse
   d. Pharm.D.

9. Radiopharmaceuticals are considered
   a. hazardous materials.
   b. OTC medications.
   c. safe to store in unrestricted areas.
   d. exempt narcotics.

10. Body badges for technicians that work with radiopharmaceuticals are typically monitored
    a. weekly.
    b. monthly.
    c. quarterly.
    d. yearly.

11. Compounded prescriptions are usually covered by
    a. private insurance companies.
    b. Medicaid.
    c. Medicare.
    d. public insurance.

PHARMACY TECHNICIAN LABS

The following pages contain exercises to provide hands-on activities to accompany each of the chapters in *The Pharmacy Technician,* Sixth Edition. They are intended to provide students with opportunities for engaging experiences to enhance learning of some concepts associated with the chapters.

Supplies for labs: A pencil, a pen, a bound composition book, index cards (your instructor may have a preference for what size you should use), and a highlighter.

You will need access to the Internet to complete these labs. Web addresses current at the time of printing have been included in the exercises. If you find a printed web address is not available when you are working on the exercise, you may wish to use a different link.

You should use a bound notebook ,such as a composition book, to serve as a lab notebook for documenting your lab activities. Throughout the labs, you should make notes and write the answers to any questions in your lab notebook.

Learning activities for learning the brand names, generic names, and classifications of the top 200 brand name drugs have been separated by classification, and dispersed throughout each of the labs to make memorization of information more manageable.

DRUG CARDS

Drug cards can be a great study tool for memorizing information about drugs. It is suggested that you prepare and learn the information on your drug cards in advance of working on each lab. That way, you can build on the foundational knowledge with your lab experiences.

INSTRUCTIONS FOR MAKING DRUG CARDS

1. Use pencil to write on the index card so you can easily change or correct any information. Use highlighters to color-code your cards, giving each type of information a different color. You may wish to include additional information to customize the information for your drug cards.

2. Write the brand name of the drug on the front of the index card.

3. Write the generic name and classification on the backside of the index card. You can use a different color of highlighter for the generic name and the classification.

4. Your instructor may want you to add additional cards for commonly used generic drugs or your instructor may provide additional information for you to include on each drug card. You can use online resources such as MedlinePlus of the National Library of Medicine website, http://www.nlm.nih.gov/medlineplus/.

5. Study and memorize the information you have written on your drug cards. You can test yourself by looking at the brand names on the front and trying to remember the information on the back, or looking at the information on the back and trying to remember the brand names of the drugs.

PRESCRIPTION LABELS

Many of the lab activities will ask you to prepare prescription labels. You can use an index card formatted like the one below or prepare a computer-generated one if you have access to a pharmacy software system.

	Pharmacy Name Address Phone Number	
Rx Number	Dr.	
Patient		
Directions		
Drug		
Date	R.Ph.	______Refills

LAB ONE

ANALGESICS & MUSCULOSKELETAL AGENTS

DRUG CARDS

Make drug cards for the following drugs. (See page 209 for instructions on making drug cards.)

Brand Name	Generic Name	Classification
Celebrex	celecoxib	Analgesic, NSAID
Voltaren gel	diclofenac topical	Analgesic, NSAID
Arthrotec	diclofenac and misoprostol	Analgesic, NSAID
OxyContin	oxycodone	Analgesic, opiate
Opana ER	oxymorphone	Analgesic, opiate
Lidoderm	lidocaine transdermal	Anesthetic, ocal
Skelaxin	metaxalone	Musculoskeletal, Muscle relaxant
Boniva	ibandronate	Musculoskeletal, Osteoporitics
Evista	raloxifene	Musculoskeletal, Osteoporitics
Actonel	risedronate	Musculoskeletal, Osteoporitics
Actonel 150	risedronate	Musculoskeletal, Osteoporitic

LAB ACTIVITIES

1. Visit the FDA website, www.fda.gov/, and click on the Drugs tab. What items are listed in the Spotlight box? (Write your answers in your lab notebook.)

2. Visit the website for QS/1® http://qs1.com/ and click on the link for NRx Pharmacy Management link. Once at the the NRx Pharmacy Management site, click on the View Demo link, and watch the demo. List the activities from the Computers in Pharmacy spread of *The Pharmacy Technician,* Sixth Edition text (pages 14–15) that are provided by the NRx Pharmacy Management software.

3. Visit the website for the American Society of Health-System Pharmacists at www.ashp.org/, and click on the link for Accreditation. Once in the accreditation website, click on the link for Technician Training Directory. Search for the accredited technician training programs in your state and write the names of the programs in your lab notebook.

4. Visit the Pharmacy Technician Certification Board website, www.ptcb.org/, and click on the tab for Certification. Click on the link for Candidate Guidebook for the most current information about the PTCE exam.

5. Visit www.nhanow.com/pharmacy-technician.aspx, the NHA website for the ExCPT exam. Click on the link for the Pharmacy Technician Candidate Handbook for the most current information about the ExCPT exam.

6. Visit the MedlinePlus website of the National Library of Medicine, http://www.nlm.nih.gov/medlineplus/.

a. Search MedlinePlus for Oxycontin® and select the druginfo result for oxycodone (http://www.nlm.nih.gov/medlineplus/druginfo/meds/a682132.htm) to answer the following questions:

1. Why is this medication prescribed?
2. What storage conditions are needed for this medicine?
3. What are other brand names for oxycodone products?
4. What are brand names for combination products containing oxycodone and what does each product contain?
5. Read the pronunciation of oxycodone at the top of the page and correctly pronounce oxycodone.

b. Search MedlinePlus for Celebrex® and select the druginfo result for celecoxib (http://www.nlm.nih.gov/medlineplus/druginfo/meds/a699022.html) to answer the following questions:

1. Why is this medication prescribed?
2. What storage conditions are needed for this medicine?
3 Are there any other brand names for celecoxib products?
4. Read the pronunciation of celecoxib at the top of the page and correctly pronounce celecoxib.

c. Search MedlinePlus for Boniva® and select the druginfo result for ibandronate (http://www.nlm.nih.gov/medlineplus/druginfo/meds/a605035.html) to answer the following questions:

1. Why is this medication prescribed?
2. What storage conditions are needed for this medicine?
3. Are there any other brand names for ibandronate products?
4. Read the pronunciation of ibandronate at the top of the page and correctly pronounce ibandronate.

LAB TWO

ANTI-INFECTIVES

DRUG CARDS

Make drug cards for the following drugs. (See page 209 for instructions on making drug cards.)

Brand Name	Generic Name	Classification
Levaquin	levofloxacin	Anti-Infective
Avelox	moxifloxacin	Anti-Infective
Solodyn	minocycline	Anti-Infective
Doryx	doxycycline	Anti-Infective
Valtrex	valacyclovir	Anti-Infective, Antiviral
Tamiflu	oseltamivir	Anti-Infective, Antiviral
Truvada	tenofovir/emtricitabine	Anti-Infective, Antiviral
Norvir	ritonavir	Anti-Infective, Antiviral
Femara	letrozole	Antineoplastic, Hormone
Arimidex	anastrozole	Antineoplastic, Hormone

LAB ACTIVITIES

1. Visit the FDA video gallery website for drugs at www.accessdata.fda.gov/scripts/video/drugs.cfm, and select two videos to watch. Write two sentence summaries about the videos in your lab notebook.

2. Visit the FDA website for Drug Safety Communications, www.fda.gov/Drugs/DrugSafety/ucm199082.htm. List two current drug safety communications.

3. Visit the DEA website for the Pharmacist's Manual at www.deadiversion.usdoj.gov/pubs/manuals/pharm2/index.html. Click on the link for Section VIII – Ordering Controlled Substances and explain how additional DEA Form 222 order forms can be requested from the DEA.

4. Visit the website for the National Association of Boards of Pharmacy, www.nabp.net/, and click on the link for Boards of Pharmacy. Then click on the link for your state (if in the United States) or province (if in Canada) and write down the website address for your board of pharmacy.

5. Visit the website for The Joint Commission, at www.jointcommission.org/, and click on the Accreditation link. List three types of health-care organizations that are accredited by TJC.

6. Visit the MedlinePlus website of the National Library of Medicine, http://www.nlm.nih.gov/medlineplus/.

a. Search MedlinePlus for Levaquin® and select the druginfo result for levofloxacin (http://www.nlm.nih.gov/medlineplus/druginfo/meds/a697040.html) to answer the following questions:

1. Why is this medication prescribed?
2. What storage conditions are needed for this medicine?
3. Are there any special dietary instructions? What auxiliary label could be used for this dietary instruction?
4. Read the pronunciation of levofloxacin at the top of the page and correctly pronounce levofloxacin.

b. Search MedlinePlus for Valtrex® and select the druginfo result for valacyclovir (http://www.nlm.nih.gov/medlineplus/druginfo/meds/a695010.html) to answer the following questions:

1. Why is this medication prescribed?
2. What storage conditions are needed for this medicine?
3. Are there any other brand names for valacyclovir products?
4. Read the pronunciation of valacyclovir at the top of the page and correctly pronounce valacyclovir.

c. Search MedlinePlus for Tamiflu® and select the druginfo result for oseltamivir (http://www.nlm.nih.gov/medlineplus/druginfo/meds/a699040.html) to answer the following questions:

1. Why is this medication prescribed?
2. How should the suspension be used?
3. What storage conditions are needed for this medicine?
4. Are there any other brand names for oseltamivir products?
5. Read the pronunciation of oseltamivir at the top of the page and correctly pronounce oseltamivir.

LAB THREE

CARDIOVASCULAR AGENTS I

DRUG CARDS

Make drug cards for the following drugs. (See page 209 for instructions on making drug cards.)

Brand Name	Generic Name	Classification
Diovan	valsartan	Cardiovascular, Antihypertensive
Diovan HCT	hydrochlorothiazide and valsartan	Cardiovascular, Antihypertensive
Benicar	olmesartan	Cardiovascular, Antihypertensive
Benicar HCT	olmesartan and hydrochlorothiazide	Cardiovascular, Antihypertensive
Avapro	irbesartan	Cardiovascular, Antihypertensive
Bystolic	nebivolol	Cardiovascular, Antihypertensive
Avalide	irbesartan and hydrochlorothiazide	Cardiovascular, Antihypertensive
Cozaar	losartan	Cardiovascular, Antihypertensive
Micardis	telmisartan	Cardiovascular, Antihypertensive
Exforge	amlodipine and valsartan	Cardiovascular, Antihypertensive
Micardis HCT	telmisartan and hydrochlorothiazide	Cardiovascular, Antihypertensive
Lotrel	amlodipine and benazepril	Cardiovascular, Antihypertensive
Hyzaar	losartan and hydrochlorothiazide	Cardiovascular, Antihypertensive
Coreg CR	carvedilol	Cardiovascular, Antihypertensive
Toprol XL	metoprolol	Cardiovascular, Antihypertensive
Azor	amlodipine and olmesartan	Cardiovascular, Antihypertensive
Tekturna	aliskiren	Cardiovascular, Antihypertensive
Atacand	candesartan	Cardiovascular, Antihypertensive

LAB ACTIVITIES

1. Visit the website for the National Library of Medicine, www.nlm.nih.gov/, and click on the link for MedlinePlus. Search the medical dictionary feature for the following terms and write the definitions in your lab notebook: hypoglycemia, hyperthyroidism, cardiomyopathy, and sublingual.

2. Visit the website for the National Library of Medicine, www.nlm.nih.gov/, and click on the link for MedlinePlus. Then Click on the link for the Medical Encyclopedia. Find and watch the videos on the following organ systems:

a. cardiovascular
b. endocrine glands
c. lymphatics

3. Visit the MedlinePlus website of the National Library of Medicine http://www.nlm.nih.gov/medlineplus/.

a. Search MedlinePlus for Benicar® and select the druginfo result for olmesartan, (http://www.nlm.nih.gov/medlineplus/druginfo/meds/a603006.htm) to answer the following questions:

1. Why is this medication prescribed?
2. What storage conditions are needed for this medicine?
3. What are other brand names for olmesartan products (including combination products)?
4. Read the pronunciation of olmesartan at the top of the page and correctly pronounce olmesartan.

b. Search MedlinePlus for Benicar® HCT and select the druginfo result for olmesartan and hydrochlorothiazide, (http://www.nlm.nih.gov/medlineplus/druginfo/meds/a611030.html) to answer the following questions:

1. What two medications are contained in this product?
2. Read the pronunciation of olmesartan and hydrochlorothiazide at the top of the page and correctly pronounce olmesartan and hydrochlorothiazide.
3. Click on the link for hydrochlorothiazide. Why is hydrochlorothiazide prescribed?

c. Search MedlinePlus for Toprol® XL and select the druginfo result for metoprolol, (http://www.nlm.nih.gov/medlineplus/druginfo/meds/a682864.html) to answer the following questions:

1. Why is this medication prescribed? What is noted about extended-release metoprolol?
2. What storage conditions are needed for this medicine?
3. Are there any other brand names for metoprolol products? What other products contain metoprolol?
4. Read the pronunciation of metoprolol at the top of the page and correctly pronounce metoprolol.

LAB FOUR

CARDIOVASCULAR AGENTS II

DRUG CARDS

Make drug cards for the following drugs. (See page 209 for instructions on making drug cards.)

Brand Name	Generic Name	Classification
Plavix	clopidogrel	Cardiovascular
Aggrenox	aspirin and dipyridamole	Cardiovascular
Nitrostat	nitroglycerin	Cardiovascular
Aspir-Low	aspirin and dipyridamole	Cardiovascular
Ranexa	ranolazine	Cardiovascular, Antiangina
Lanoxin	digoxin	Cardiovascular, Antiarrhythmic
Coumadin	warfarin	Cardiovascular, Anticoagulant
Lovenox	enoxaparin	Cardiovascular, Anticoagulant
Lipitor	atorvastatin	Cardiovascular, Antihyperlipidemic
Crestor	rosuvastatin	Cardiovascular, Antihyperlipidemic
Tricor	fenofibrate	Cardiovascular, Antihyperlipidemic
Zetia	ezetimibe	Cardiovascular, Antihyperlipidemic
Vytorin	ezetimibe and simvastatin	Cardiovascular, Antihyperlipidemic
Niaspan	niacin	Cardiovascular, Antihyperlipidemic
Lovaza	omega-3 fatty acid	Cardiovascular, Antihyperlipidemic
Trilipix	fenofibrate	Cardiovascular, Antihyperlipidemic
Welchol	colesevelam	Cardiovascular, Antihyperlipidemic
Simcor	niacin	Cardiovascular, Antihyperlipidemic
Caduet	amlodipine and atorvastatin	Cardiovascular, Antihyperlipidemic, Antihypertensive

LAB ACTIVITIES

1. Visit the website for the Institute for Safe Medication Practices, www.ismp.org/, and click on the link for High-Alert Medications Consumer Leaflets. List 10 medications that have these leaflets. Print the leaflet for warfarin. List the colors of warfarin tablets that are associated with each strength.

2. Visit the website for the Institute for Safe Medication Practices at www.ismp.org, and click on the link for Tools. Then Click on the link for the Error-Prone Abbreviations List and print the list. Write the intended meaning for the following abbreviations in your lab notebook and the possible misinterpretations: AD, AS, AU, OD, OS, OU, OD, OS, OU. Also, indicate the better way to communicate the intended meaning.

3. Visit the Medline Plus website of the National Library of Medicine, http://www.nlm.nih.gov/medlineplus/. Click on the tab for Drugs & Supplements, then "browse" for drug information on digoxin oral (Lanoxin®), warfarin (Coumadin®), and omega-3-acid ethyl esters (Lovaza®). Read the drug information page for each drug and answer the following question.

a. Why are these medications prescribed?
b. What storage conditions are needed for these medicines?
c. What are other brand names for digoxin and for warfarin?
d. Are there brand name combination products containing this drug and if so what does each product contain?
e. What is the special precaution noted about aspirin or aspirin-containing products when using warfarin? What auxiliary label could be used for this precaution?
f. Read the drug pronunciation at the top of the page; correctly pronounce it.

4. Prepare a label for each of the following prescriptions using an index card formatted like the example shown on page 209, or prepare a computer-generated label if a computer pharmacy software system is available. Assume you have checked the patient's profile and found that the patient is not taking any other medications, has no medication allergies, and pays cash for the prescriptions.

a. A prescription was written by Dr. Alice Chan for Andy Apple (date of birth 7/14/1953) for Lanoxin® 0.25 mg #30, Sig: i tab daily, 2 refills.

b. A prescription was written by Dr. Alice Chan for Bobby Banana (date of birth 2/24/1948) for Coumadin® 5 mg #30, Sig: i tab daily, No refills.

c. A prescription was written by Dr. Alice Chan for Carol Crumb (date of birth 9/4/1943) for Lovaza® 1 g #120, Sig: ii caps b.i.d., 2 refills.

LAB FIVE

DENTAL & DERMATOLOGICAL AGENTS

DRUG CARDS

Make drug cards for the following drugs. (See page 209 for instructions on making drug cards.)

Brand Name	Generic Name	Classification
SF 5000 Plus	fluoride	Dental
Zovirax topical	acyclovir	Dermatological
Differin	adapalene	Dermatological
Bactroban	mupirocin	Dermatological
BenzaClin	clindamycin and benzoyl peroxide	Dermatological
Duac Care System	clindamycin and benzoyl peroxide	Dermatological
Epiduo	adapalene and benzoyl peroxide	Dermatological

LAB ACTIVITIES

1. Visit the MedlinePlus website of the National Library of Medicine, http://www.nlm.nih.gov/medlineplus/.

    a. Search MedlinePlus for SF 5000 Plus® and select the druginfo result for fluoride (http://www.nlm.nih.gov/medlineplus/druginfo/meds/a682727.html) to answer the following questions:

    1. Why is this medication prescribed?
    2. What storage conditions are needed for this medicine?
    3. What are other brand names for fluoride products?
    4. Read the pronunciation of fluoride at the top of the page and correctly pronounce fluoride.

b. Search MedlinePlus for Differin® and select the druginfo result for adapalene (http://www.nlm.nih.gov/medlineplus/druginfo/meds/a604001.html) to answer the following questions:

1. Why is this medication prescribed?
2. What storage conditions are needed for this medicine?
3 What are the special precautions for adapalene products?
4. Read the pronunciation of adapalene at the top of the page and correctly pronounce adapalene.

c. Search MedlinePlus for Benzaclin® and select the druginfo result for clindamycin and benzoyl peroxide topical (http://www.nlm.nih.gov/medlineplus/druginfo/meds/a603021.html) to answer the following questions:

1. Why is this medication prescribed?
2. What storage conditions are needed for this medicine?
3. What are other brand names for clindamycin and benzoyl peroxide products?
4. What are the special precautions for clindamycin and benzoyl peroxide products?
5. Read the pronunciation of clindamycin and benzoyl peroxide at the top of the page and correctly pronounce clindamycin and benzoyl peroxide.

2. Prepare a label for each of the following prescriptions using an index card formatted like the example shown on page 209, or prepare a computer-generated label if a computer pharmacy software system is available. Assume you have checked the patient's profile and found that the patient is not taking any other medications, has no medication allergies, and pays cash for the prescriptions.

a. A prescription was written by Dr. Alice Chan for David Noodle (date of birth 7/14/1953) for SF 5000 Plus® Disp. 51 g, Sig: Apply a thin ribbon of SF 5000 Plus to toothbrush. Brush thoroughly for two minutes, preferably at bedtime, 3 refills.

b. A prescription was written by Dr. Alice Chan for Sara Sunny (date of birth 2/24/1948) for Differin® 0.1% Cream Disp. 45 g, Sig: daily hs. 1 refill.

c. A prescription was written by Dr. Alice Chan for Tom Tree (date of birth 9/4/1943) for Benzaclin® Disp. 25 g, Sig: apply b.i.d., 2 refills.

LAB SIX

GASTROINTESTINAL AGENTS

DRUG CARDS

Make drug cards for the following drugs. (See page 209 for instructions on making drug cards.)

Brand Name	Generic Name	Classification
Moviprep	polyethylene glycol electrolyte	Gastrointentinal
Dexilant/Kapidex	dexlansoprazole	Gastrointestinal
Asacol	mesalamine	Gastrointestinal
Amitiza	lubiprostone	Gastrointestinal
Transderm-Scop	scopolamine	Gastrointestinal
Nexium	esomeprazole	Gastrointestinal, Antacid/Antiulcer
Aciphex	rabeprazole	Gastrointestinal, Antacid/Antiulcer
Protonix	pantoprazole	Gastrointestinal, Antacid/Antiulcer
Prevacid SoluTab	lansoprazole	Gastrointestinal, Antacid/Antiulcer
Prevacid	lansoprazole	Gastrointestinal, Antacid/Antiulcer
Carafate	sucralfate	Gastrointestinal, Antacid/Antiulcer
HalfLytely bowel prep	PEG-3350, sodium chloride, sodium bicarbonate and potassium chloride	Gastrointestinal, Laxative

LAB ACTIVITIES

1. Visit the MedlinePlus website of the National Library of Medicine, http://www.nlm.nih.gov/medlineplus/.

   a. Search MedlinePlus for polyethylene glycol electrolyte solution and select the druginfo result for polyethylene glycol electrolyte solution (http://www.nlm.nih.gov/medlineplus/druginfo/meds/a601097.html) to answer the following questions:

   1. Why is this medication prescribed?
   2. What storage conditions are needed for this medicine?
   3. What are other brand names for polyethylene glycol electrolyte solution products?
   4. Read the pronunciation of polyethylene glycol at the top of the page and correctly pronounce polyethylene glycol.

b. Search MedlinePlus for Transderm-Scop® and select the druginfo result for scopolamine patch (http://www.nlm.nih.gov/medlineplus/druginfo/meds/a682509.html) to answer the following questions:

1. Why is this medication prescribed?
2. What storage conditions are needed for this medicine?
3. What are the special precautions for scopolamine patch products?
4. Read the pronunciation of scopolamine at the top of the page and correctly pronounce scopolamine.

c. Search MedlinePlus for Prevacid® Solutab and select the druginfo result for lansoprazole (http://www.nlm.nih.gov/medlineplus/druginfo/meds/a695020.html) to answer the following questions:

1. Why is this medication prescribed?
2. What storage conditions are needed for this medicine?
3 What are other brand names for lansoprazole products?
4. What are the special precautions for lansoprazole products?
5. Read the pronunciation of lansoprazole at the top of the page and correctly pronounce lansoprazole.

2. Prepare a label for each of the following prescriptions using an index card formatted like the example shown on page 209, or prepare a computer-generated label if a computer pharmacy software system is available. Assume you have checked the patient's profile and found that the patient is not taking any other medications, has no medication allergies, and pays cash for the prescriptions.

a. A prescription was written by Dr. Alice Chan for Derrick Dimple (date of birth 7/14/1953) for Moviprep Disp 2 liters, Sig: 8 oz q15 min x4 doses, repeat 90 min later, then drink 1 l of clear liquid evening before procedure, no refills.

b. A prescription was written by Dr. Alice Chan for George Grace (date of birth 2/24/1948) for Transderm-Scop #4, Sig: Start: evening prior to surgery. Info: remove patch 24h after surgery; do not cut patch. No refills.

c. A prescription was written by Dr. Alice Chan for Cindy Current (date of birth 9/4/1943) for Prevacid Solutab 30 mg #30, Sig: i PO daily, before breakfast, 2 refills.

LAB SEVEN

HORMONES & MODIFIERS I

DRUG CARDS

Make drug cards for the following drugs. (See page 209 for instructions on making drug cards.)

Brand Name	Generic Name	Classification
Synthroid	levothyroxine	Hormones & Modifiers, Thyroid
Thyroid, Armour	dessicated thyroid	Hormones & Modifiers, Thyroid
Prempro	conjugated estrogens and medroxyprogesterone	Hormones & Modifiers
Propecia	finasteride	Hormones & Modifiers
AndroGel	testosterone	Hormones & Modifiers, Androgen
Premarin tabs	conjugated estrogens	Hormones & Modifiers, Estrogen
Vivelle-DOT	estradiol	Hormones & Modifiers, Estrogen
Vagifem	estrogen	Hormones & Modifiers, Estrogen
Premarin vaginal	conjugated estrogens	Hormones & Modifiers, Estrogen
Estrace vaginal	estradiol	Hormones & Modifiers, Estrogen
Prometrium	progesterone	Hormones & Modifiers, Progestins
Levoxyl	levothyroxine	Hormones & Modifiers, Thyroid
Levothroid	levothyroxine	Hormones & Modifiers, Thyroid

LAB ACTIVITIES

1. Visit the MedlinePlus website of the National Library of Medicine, http://www.nlm.nih.gov/medlineplus/.

   a. Search MedlinePlus for Synthroid® and select the druginfo result for levothyroxine (http://www.nlm.nih.gov/medlineplus/druginfo/meds/a682461.html) to answer the following questions:

   1. Why is this medication prescribed?
   2. What storage conditions are needed for this medicine?
   3. What are other brand names for levothyroxine products?
   4. Read the pronunciation of levothyroxine at the top of the page and correctly pronounce levothyroxine.

b. Search MedlinePlus for Propecia® and select the druginfo result for finasteride (http://www.nlm.nih.gov/medlineplus/druginfo/meds/a698016.html) to answer the following questions:

1. Why is this medication prescribed?
2. What storage conditions are needed for this medicine?
3. What is another brand name for a finasteride product?
4. What are the special precautions for finasteride products?
5. Read the pronunciation of finasteride at the top of the page and correctly pronounce finasteride.

c. Search MedlinePlus for Androgel® and select the druginfo result for testosterone topical (http://www.nlm.nih.gov/medlineplus/druginfo/meds/a605020.html) to answer the following questions:

1. What is the important warning in a red box for this medication?
2. Why is this medication prescribed?
3. What storage conditions are needed for this medicine?
4. What is another brand name for a testosterone topical product?
5. What are the special precautions for testosterone products?
6. Read the pronunciation of testosterone at the top of the page and correctly pronounce testosterone.

2. Prepare a label for each of the following prescriptions using an index card formatted like the example shown on page 209, or prepare a computer-generated label if a computer pharmacy software system is available. Assume you have checked the patient's profile and found that the patient is not taking any other medications, has no medication allergies, and pays cash for the prescriptions.

a. A prescription was written by Dr. Alice Chan for Donna Dimple (date of birth 7/14/1953) for Synthroid® 0.1 mg DAW #30, Sig: i tab daily, 2 refills.

b. A prescription was written by Dr. Alice Chan for Eddy Elmer (date of birth 2/24/1948) for Propecia® 1mg #30, Sig: i tab daily, 2 refills.

c. A prescription was written by Dr. Alice Chan for Curt Crumb (date of birth 9/4/1943) for Androgel® 1% Packets #30, Sig: i packet daily, 2 refills. When filling the prescription you notice the symbol C-III on the package. What does the C-III mean?

LAB EIGHT

HORMONES & MODIFIERS II

DRUG CARDS

Make drug cards for the following drugs. (See page 209 for instructions on making drug cards.)

Brand Name	Generic Name	Classification
Lantus	insulin glargine	Hormones & Modifiers, Insulin
NovoLog	insulin aspart (rDNA origin)	Hormones & Modifiers, Insulin
Lantus SoloSTAR	insulin glargine (rDNA origin)	Hormones & Modifiers, Insulin
Humalog	insulin lispro	Hormones & Modifiers, Insulin
Levemir	insulin detemir (rDNA origin)	Hormones & Modifiers, Insulin
Humulin N	insulin (human recombinant)	Hormones & Modifiers, Insulin
NovoLog Mix 70/30	insulin aspart protamine and insulin aspart (rDNA origin)	Hormones & Modifiers, Insulin
Humulin 70/30	insulin (human recombinant)	Hormones & Modifiers, Insulin
Novolin 70/30	human insulin isophane suspension and regular, human insulin injection (rDNA origin)	Hormones & Modifiers, Insulin
Humulin R	insulin (human recombinant)	Hormones & Modifiers, Insulin
Humalog Mix 75/25 Pen	insulin lispro protamine suspension mixed with soluble insulin lispro	Hormones & Modifiers, Insulin
Byetta	exenatide	Hormones & Modifiers

LAB ACTIVITIES

1. Visit the MedlinePlus website of the National Library of Medicine, http://www.nlm.nih.gov/medlineplus/.

   a. Search MedlinePlus for Lantus® and select the druginfo result for insulin glargine (rDNA origin) injection (http://www.nlm.nih.gov/medlineplus/druginfo/meds/a600027.html) to answer the following questions:

   1. Why is this medication prescribed?
   2. What storage conditions are needed for this medicine?
   3. Read the pronunciation of insulin glargine at the top of the page and correctly pronounce insulin glargine.

b. Search MedlinePlus for Levemir® and select the druginfo result for insulin detemir (rDNA Origin) injection (http://www.nlm.nih.gov/medlineplus/druginfo/meds/a606012.html) to answer the following questions:
1. Why is this medication prescribed?
2. What storage conditions are needed for this medicine?
3. Read the pronunciation of insulin detemir at the top of the page and correctly pronounce insulin detemir.

c. Search MedlinePlus for Byetta® and select the druginfo result for exenatide injection (http://www.nlm.nih.gov/medlineplus/druginfo/meds/a605034.html) to answer the following questions:
1. Why is this medication prescribed?
2. What storage conditions are needed for this medicine?
3. What are the special precautions for exenatide products?
4. Read the pronunciation of exenatide at the top of the page and correctly pronounce exenatide.
5. Follow the link to the FDA Medication Guide for Byetta®.

2. Prepare a label for each of the following prescriptions using an index card formatted like the example shown on page 209, or prepare a computer-generated label if a computer pharmacy software system is available. Assume you have checked the patient's profile and found that the patient is not taking any other medications, has no medication allergies, and pays cash for the prescriptions.

a. A prescription was written by Dr. Alice Chan for David Noodle (date of birth 7/14/1953) for Lantus® SoloSTAR 100 Units/ml Disp. #5 g, Sig: 10 units SC qhs, 8 refills. Another prescription was written for Pen Needles Disp #60, Sig: use b.i.d. for Lantus® injections. 8 refills.

b. A prescription was written by Dr. Alice Chan for Sara Sunny (date of birth 2/24/1948) for Levemir® 100 Units/ml Disp 10 ml, Sig: 10 units SC b.i.d. 5 refills. Another prescription was written for Insulin syringes and needles 3/10 cc Disp #60, Sig: use b.i.d. for insulin injections. 5 refills.

c. A prescription was written by Dr. Alice Chan for Tom Tree (date of birth 9/4/1943) for Byetta® 5 mcg/0.02 ml Disp. 1 pen, Sig 5 mcg SC b.i.d. within 1h before am and pm meals. 2 refills. Another prescription was written for Pen Needles Disp #60, Sig: use b.i.d. for Byetta® injections. 2 refills.

LAB NINE

HORMONES & MODIFIERS III

DRUG CARDS

Make drug cards for the following drugs. (See page 209 for instructions on making drug cards.)

Brand Name	Generic Name	Classification
Actos	pioglitazone	Hormones & Modifiers, Oral antidiabetic
Januvia	sitagliptin	Hormones & Modifiers, Oral antidiabetic
Glipizide XL	glipizide	Hormones & Modifiers, Oral antidiabetic
Janumet	metformin and sitagliptin	Hormones & Modifiers, Oral antidiabetic
Actoplus Met	metformin and pioglitazone	Hormones & Modifiers, Oral antidiabetic
Avandia	rosiglitazone	Hormones & Modifiers, Oral antidiabetic
Prandin	repaglinide	Hormones & Modifiers, Oral antidiabetic

LAB ACTIVITIES

1. Visit the MedlinePlus website of the National Library of Medicine http://www.nlm.nih.gov/medlineplus/.

   a. Search MedlinePlus for Actos® and select the druginfo result for pioglitazone (http://www.nlm.nih.gov/medlineplus/druginfo/meds/a699016.html) to answer the following questions:

   1. Why is this medication prescribed?
   2. What storage conditions are needed for this medicine?
   3. What are other brand names for pioglitazone products?
   4. Read the pronunciation of pioglitazone at the top of the page and correctly pronounce pioglitazone.

   b. Search MedlinePlus for Januvia® and select the druginfo result for sitagliptin (http://www.nlm.nih.gov/medlineplus/druginfo/meds/a606023.html) to answer the following questions:

   1. Why is this medication prescribed?
   2. What storage conditions are needed for this medicine?
   3. What are other brand names for sitagliptin products?
   4. What are the special precautions for sitagliptin products?
   5. Read the pronunciation of sitagliptin at the top of the page and correctly pronounce sitagliptin.

c. Search MedlinePlus for Prandin® and select the druginfo result for repaglinide (http://www.nlm.nih.gov/medlineplus/druginfo/meds/a600010.html) to answer the following questions:

1. Why is this medication prescribed?
2. What storage conditions are needed for this medicine?
3. What is the brand name of another repaglinide product?
4. What are the special precautions for repaglinide products?
5. Read the pronunciation of repaglinide at the top of the page and correctly pronounce repaglinide.

2. Prepare a label for each of the following prescriptions using an index card formatted like the example shown on page 209, or prepare a computer-generated label if a computer pharmacy software system is available. Assume you have checked the patient's profile and found that the patient is not taking any other medications, has no medication allergies, and pays cash for the prescriptions.

a. A prescription was written by Dr. Alice Chan for Donna Green (date of birth 7/14/1953) for Actos® 15 mg Disp. #30 g, Sig: i daily, 3 refills. .

b. A prescription was written by Dr. Alice Chan for Susan Knight (date of birth 2/24/1948) for Januvia® 100 mg Disp. #30, Sig: i daily a.m., 5 refills.

c. A prescription was written by Dr. Alice Chan for Tony Green (date of birth 9/4/1943) for Prandin® 0.5 mg Disp. #90, Sig: i a.c., 3 refills.

LAB TEN

HORMONES & MODIFIERS IV

DRUG CARDS

Make drug cards for the following drugs. (See page 209 for instructions on making drug cards.)

Brand Name	Generic Name	Classification
Loestrin 24 Fe	ethinyl estradiol and norethindrone and iron	Hormones & Modifiers, Contraceptive
NuvaRing	etonogestrel and ethinyl estradiol	Hormones & Modifiers, Contraceptive
Yaz	drosperinone and ethinyl estradiol	Hormones & Modifiers, Contraceptive
Ortho Tri-Cyclen Lo	norgestimate and ethinyl estradiol	Hormones & Modifiers, Contraceptive
Apri	desogestrel and ethinyl estradiol	Hormones & Modifiers, Contraceptive
Kariva	desogestrel and ethinyl estradiol	Hormones & Modifiers, Contraceptive
Ortho Evra	norelgestromin and ethinyl estradiol	Hormones & Modifiers, Contraceptive
Yasmin 28	drospirenone and ethinyl estradiol	Hormones & Modifiers, Contraceptive
Ortho Tri-Cyclen	norgestimate and ethinyl estradiol	Hormones & Modifiers, Contraceptive

LAB ACTIVITIES

1. Visit the MedlinePlus website of the National Library of Medicine, http://www.nlm.nih.gov/medlineplus/.

   a. Search MedlinePlus for Loestrin® 24 Fe and select the druginfo result for estrogen and progestin (oral contraceptives) (http://www.nlm.nih.gov/medlineplus/druginfo/meds/a601050.html) to answer the following questions:

   1. Why is this medication prescribed?
   2. What storage conditions are needed for this medicine?
   3. What are five other brand names for oral contraceptive products?

b. Search MedlinePlus for Nuvaring® and select the druginfo result for ethinyl estradiol and etonogestrel vaginal ring (http://www.nlm.nih.gov/medlineplus/druginfo/meds/a604032.html) to answer the following questions:

1. Why is this medication prescribed?
2. What storage conditions are needed for this medicine?
3. What are the special precautions for ethinyl estradiol and etonogestrel vaginal ring products?
4. What is a special dietary instruction for this product?
5. Read the pronunciation of ethinyl estradiol and etonogestrel at the top of the page and correctly pronounce ethinyl estradiol and etonogestrel.

c. Search MedlinePlus for Ortho Evra®and select the druginfo result for ethinyl estradiol and norelgestromin transdermal (http://www.nlm.nih.gov/medlineplus/druginfo/meds/a602006.html) to answer the following questions:

1. Why is this medication prescribed?
2. What storage conditions are needed for this medicine?
3 What are the special precautions for ethinyl estradiol and norelgestromin transdermal products?
4. Read the pronunciation of ethinyl estradiol and norelgestromin at the top of the page and correctly pronounce ethinyl estradiol and norelgestromin.

2. Prepare a label for each of the following prescriptions using an index card formatted like the example shown on page 209, or prepare a computer-generated label if a computer pharmacy software system is available. Assume you have checked the patient's profile and found that the patient is not taking any other medications, has no medication allergies, and pays cash for the prescriptions.

a. A prescription was written by Dr. Alice Chan for Debbie Downey (date of birth 7/14/1953) for Loestrin® 24 Fe Disp. #28 g, Sig: i daily, 11 refills.

b. A prescription was written by Dr. Alice Chan for Susan Knight (date of birth 2/24/1948) for Nuvaring® Disp. #1, Sig: 1 ring PV x3wk, off x1wk. 11 refills.

c. A prescription was written by Dr. Alice Chan for Tonia Simpson (date of birth 9/4/1943) for Ortho Evra® Disp. #1 box, Sig: apply 1 patch qwk x3wk, off x1wk, 11 refills.

LAB ELEVEN

NEUROLOGICAL & PSYCHOTROPIC AGENTS

DRUG CARDS

Make drug cards for the following drugs. *(See page 209 for instructions on making drug cards.)*

Brand Name	Generic Name	Classification
Lyrica	pregabalin	Neurological
Aricept	donepezil	Neurological, Anti-Alzheimer's
Namenda	memantine	Neurological, Anti-Alzheimer's
Exelon Patch	rivastigmine	Neurological, Anti-Alzheimer's
Lamictal	lamotrigine	Neurological, Antiepileptic
Dilantin	phenytoin	Neurological, Antiepileptic
Relpax	eletriptan	Neurological, Antimigraine
Maxalt	rizatriptan	Neurological, Antimigraine
Maxalt MLT	rizatriptan	Neurological, Antimigraine
Concerta	methylphenidate	Psychotropic–Neurological, ADHD
Vyvanse	lisdexamfetamine	Psychotropic–Neurological, ADHD
Adderall XR	dextroamphetamine and amphetamine	Psychotropic–Neurological, ADHD
Focalin XR	dexmethylphenidate	Psychotropic–Neurological, ADHD
Strattera	atomoxetine	Psychotropic–Neurological, ADHD
Metadate CD	methylphenidate	Psychotropic–Neurological, ADHD

LAB ACTIVITIES

1. Visit the MedlinePlus website of the National Library of Medicine http://www.nlm.nih.gov/medlineplus/.

    a. Search MedlinePlus for Exelon® Patch and select the druginfo result for rivastigmine transdermal (http://www.nlm.nih.gov/medlineplus/druginfo/meds/a607078.html) to answer the following questions:

    1. Why is this medication prescribed?
    2. What storage conditions are needed for this medicine?
    3. Read the pronunciation of rivastigmine transdermal at the top of the page and correctly pronounce rivastigmine transdermal.

b. Search MedlinePlus for Relpax® and select the druginfo result for eletriptan (http://www.nlm.nih.gov/medlineplus/druginfo/meds/a603029.html) to answer the following questions:

1. Why is this medication prescribed?
2. What storage conditions are needed for this medicine?
3. What are the special precautions for eletriptan products?
4. Read the pronunciation of eletriptan at the top of the page and correctly pronounce eletriptan.

c. Search MedlinePlus for Strattera® and select the druginfo result for atomoxetine (http://www.nlm.nih.gov/medlineplus/druginfo/meds/a603013.html) to answer the following questions:

1. Why is this medication prescribed?
2. What storage conditions are needed for this medicine?
3. What are the special precautions for atomoxetine products?
4. Read the pronunciation of atomoxetine at the top of the page and correctly pronounce atomoxetine.

2. Prepare a label for each of the following prescriptions using an index card formatted like the example shown on page 209, or prepare a computer-generated label if a computer pharmacy software system is available. Assume you have checked the patient's profile and found that the patient is not taking any other medications, has no medication allergies, and pays cash for the prescriptions.

a. A prescription was written by Dr. Alice Chan for Donna Green (date of birth 7/14/1933) for Exelon® Patch 4.6 mg Disp. #30 g, Sig: i daily, 3 refills.

b. A prescription was written by Dr. Alice Chan for Steve Day (date of birth 2/24/1948) for Relpax® 40 mg Disp. #6 Sig: i at onset, may repeat X 1 after 2 hours as directed. 1 refill.

c. A prescription was written by Dr. Alice Chan for Terry Stevens (date of birth 9/4/2001) for Strattera® 80 mg Disp. #30, Sig: i daily. No refills.

LAB TWELVE

OPHTHALMIC & OTIC AGENTS

DRUG CARDS

Make drug cards for the following drugs. *(See page 209 for instructions on making drug cards.)*

Brand Name	Generic Name	Classification
Vigamox	moxifloxacin ophthalmic	Ophthalmic
Restasis	cyclosporine	Ophthalmic
Patanol	olopatadine	Ophthalmic
Pataday	olopatadine	Ophthalmic
Zymar	gatifloxacin	Ophthalmic
Lotemax	loteprednol	Ophthalmic
Nevanac	nepafenac	Ophthalmic
Xalatan	latanoprost	Ophthalmic, Antiglaucoma
Travatan Z	travoprost	Ophthalmic, Antiglaucoma
Lumigan	bimatoprost	Ophthalmic, Antiglaucoma
Alphagan P	brimonidine	Ophthalmic, Antiglaucoma
Combigan	brimonidine and timolol	Ophthalmic, Antiglaucoma
Ciprodex otic	ciprofloxacin and dexamethasone	Otic

LAB ACTIVITIES

1. Visit the MedlinePlus website of the National Library of Medicine, http://www.nlm.nih.gov/medlineplus/.

   a. Search MedlinePlus for Restasis® and select the druginfo result for cyclosporine ophthalmic (http://www.nlm.nih.gov/medlineplus/druginfo/meds/a604009.html) to answer the following questions:

   1. Why is this medication prescribed?
   2. What storage conditions are needed for this medicine?
   3. Read the pronunciation of cyclosporine at the top of the page and correctly pronounce cyclosporine.

b. Search MedlinePlus for Xalatan® and select the druginfo result for latanoprost (http://www.nlm.nih.gov/medlineplus/druginfo/meds/a697003.html) to answer the following questions:
1. Why is this medication prescribed?
2. What storage conditions are needed for this medicine?
3. What are the special precautions for latanoprost products?
4. Read the pronunciation of latanoprost at the top of the page and correctly pronounce latanoprost.

c. Search MedlinePlus for Ciprodex® Otic and select the druginfo result for ciprofloxacin and dexamethasone otic (http://www.nlm.nih.gov/medlineplus/druginfo/meds/a607010.html) to answer the following questions:
1. Why is this medication prescribed?
2. How should the medicine be used?
3. What storage conditions are needed for this medicine?
4. What are the special precautions for ciprofloxacin and dexamethasone otic products?
5. Read the pronunciation of ciprofloxacin and dexamethasone at the top of the page and correctly pronounce ciprofloxacin and dexamethasone.

2. Prepare a label for each of the following prescriptions using an index card formatted like the example shown on page 209, or prepare a computer-generated label if a computer pharmacy software system is available. Assume you have checked the patient's profile and found that the patient is not taking any other medications, has no medication allergies, and pays cash for the prescriptions.

a. A prescription was written by Dr. Alice Chan for Sally Jones (date of birth 7/14/1953) for Restasis® 0.05%. Disp. #30 g, Sig: i gtt in each eye q 12h, 3 refills.

b. A prescription was written by Dr. Alice Chan for Jenny Smith (date of birth 2/24/1948) for Xalatan® 0.005%. Disp. 1 bottle, Sig: i gtt o.d. q.d. h.s. 5 refills.

c. A prescription was written by Dr. Alice Chan for Chad Doe (date of birth 9/4/1943) for Ciprodex® Otic. Disp. 7.5 ml, Sig: 4 gtt in both ears bid x7 days, no refills.

LAB THIRTEEN

PSYCHOTROPIC AGENTS

DRUG CARDS

Make drug cards for the following drugs. (See page 209 for instructions on making drug cards.)

Brand Name	Generic Name	Classification
Cymbalta	duloxetine	Psychotropic
Provigil	modafinil	Psychotropic
Intuniv	guanfacine extended-release	Psychotropic
Nuvigil	armodafinil	Psychotropic
Lexapro	escitalopram	Psychotropic, Antidepressant
Effexor XR	venlafaxine	Psychotropic, Antidepressant
Pristiq	desvenlafaxine	Psychotropic, Antidepressant
Wellbutrin XL	bupropion	Psychotropic, Antidepressant
Seroquel	quetiapine	Psychotropic, Antipsychotic
Abilify	aripiprazole	Psychotropic, Antipsychotic
Zyprexa	olanzapine	Psychotropic, Antipsychotic
Geodon oral	ziprasidone	Psychotropic, Antipsychotic
Seroquel XR	quetiapine	Psychotropic, Antipsychotic
Suboxone	buprenorphine and naloxone	Psychotropic, Drug dependency
Chantix	varenicline	Psychotropic, Drug dependency
Lunesta	eszopiclone	Psychotropic, Hypnotic
Ambien CR	zolpidem	Psychotropic, Hypnotic

LAB ACTIVITIES

1. Visit the MedlinePlus website of the National Library of Medicine http://www.nlm.nih.gov/medlineplus/.

   a. Search MedlinePlus for Nuvigil® and select the druginfo result for armodafinil (http://www.nlm.nih.gov/medlineplus/druginfo/meds/a607067.html) to answer the following questions:

   1. Why is this medication prescribed?
   2. What are special dietary instructions for this medicine?

3. What storage conditions are needed for this medicine?
4. Read the pronunciation of armodafinil at the top of the page and correctly pronounce armodafinil.

b. Search MedlinePlus for Suboxone® and select the druginfo result for buprenorphine sublingual (http://www.nlm.nih.gov/medlineplus/druginfo/meds/a605002.html) to answer the following questions:

1. Why is this medication prescribed?
2. What are special dietary instructions for this medicine?
3. What is the difference between Subutex® and Suboxone®?
4. What storage conditions are needed for this medicine?
5. What are special precautions for buprenorphine and naloxone products?
6. Read the pronunciation of buprenorphine at the top of the page and correctly pronounce buprenorphine.

c. Search MedlinePlus for Ambien® CR and select the druginfo result for zolpidem (http://www.nlm.nih.gov/medlineplus/druginfo/meds/a693025.html) to answer the following questions:

1. Why is this medication prescribed?
2. What is the difference between Ambien®, Ambien® CR, Edluar®, Intermezzo®, and Zolpimist®?
3. What storage conditions are needed for this medicine?
4. What are the special precautions for zolpidem products?
5. Read the pronunciation of zolpidem at the top of the page and correctly pronounce zolpidem.

2. Prepare a label for each of the following prescriptions using an index card formatted like the example shown on page 209, or prepare a computer-generated label if a computer pharmacy software system is available. Assume you have checked the patient's profile and found that the patient is not taking any other medications, has no medication allergies, and pays cash for the prescriptions.

a. A prescription was written by Dr. Alice Chan for Donna Green (date of birth 7/14/1953) for Nuvigil® 150 mg Disp. #30 g, Sig: i daily in the morning, 3 refills. When filling the prescription you notice the symbol C-IV on the package. What does the C-IV mean?

b. A prescription was written by Dr. Alice Chan for Susan Knight (date of birth 2/24/1948) for Suboxone® 2 mg /0.5 mg Disp. #30, Sig: i s.l. daily. No refills. When filling the prescription you notice the symbol C-III on the package. What does the C-III mean?

c. A prescription was written by Dr. Alice Chan for Tony Green (date of birth 9/4/1943) for Ambien® CR 12.5 mg Disp. #30, Sig: i h.s., 1 refill. When filling the prescription you notice the symbol C-IV on the package. What does the C-IV mean?

LAB FOURTEEN

RESPIRATORY AGENTS

DRUG CARDS

Make drug cards for the following drugs. (See page 209 for instructions on making drug cards.)

Brand Name	Generic Name	Classification
Singulair	montelukast	Respiratory
Advair Diskus	fluticasone and salmeterol	Respiratory
Nasonex	mometasone	Respiratory
Spiriva	tiotropium	Respiratory
Flovent HFA	fluticasone	Respiratory
Combivent	ipratropium and albuterol	Respiratory
Symbicort	formoterol and budesonide	Respiratory
Tussionex	hydrocodone and chlorpheniramine	Respiratory
Nasacort AQ	triamcinolone	Respiratory
Qvar	beclomethasone	Respiratory
Veramyst	fluticasone	Respiratory
Epipen	epinephrine	Respiratory
Asmanex	mometasone	Respiratory
Allegra-D 24 Hour	fexofenadine and pseudoephedrine	Respiratory
Advair HFA	fluticasone and salmeterol	Respiratory
Atrovent HFA	ipratropium	Respiratory
Rhinocort Aqua	budesonide	Respiratory
Pulmicort Flexhaler	budesonide	Respiratory
Omnaris	ciclesonide	Respiratory
Xyzal	levocetirizine	Respiratory, Antihistamine
Astepro 0.15%	azelastine	Respiratory, Antihistamine
Clarinex	desloratadine	Respiratory, Antihistamine
Astelin	azelastine	Respiratory, Antihistamine
ProAir HFA	albuterol	Respiratory, Bronchodilator
Ventolin HFA	albuterol	Respiratory, Bronchodilator
Proventil HFA	albuterol	Respiratory, Bronchodilator
Xopenex HFA	levalbuterol	Respiratory, Bronchodilator
Xopenex	levalbuterol	Respiratory, Bronchodilator

LAB ACTIVITIES

1. Visit the MedlinePlus website of the National Library of Medicine at www.nlm.nih.gov/medlineplus/. Click on the tab for Drugs & Supplements, then "browse" for drug information on tiotropium oral inhalation(Spiriva®), epinephrine injection (Epipen®), and albuterol inhalation (ProAir® HFA). Read the drug information page for each drug and answer the following questions.
   a. Why are these medications prescribed?
   b. What storage conditions are needed for these medicines?
   c. Read the drug pronunciation at the top of each page; correctly pronounce the drug names.
   d. What are other brand names for epinephrine and albuterol products?
   e. What are special precautions for epinephrine and albuterol products?

2. Prepare a label for each of the following prescriptions using an index card formatted like the example shown on page 209, or prepare a computer-generated label if a computer pharmacy software system is available. Assume you have checked the patient's profile and found that the patient is not taking any other medications, has no medication allergies, and pays cash for the prescriptions.

a. A prescription was written by Dr. Alice Chan for Deanna Gray (date of birth 7/14/1933) for Spiriva® 18 mcg Disp. #30 g, Sig: i cap inhaled daily, 6 refills.

b. A prescription was written by Dr. Alice Chan for Jay Day (date of birth 2/24/1948) for Epipen® Disp. #2 Sig: i IM as directed. 1 refill.

c. A prescription was written by Dr. Alice Chan for Troy Schmidt (date of birth 9/4/2001) for ProAir® HFA 8.5 g Disp. #1, Sig: 2 puffs inhaled q4–6h prn, 6 refills.

LAB FIFTEEN

URINARY TRACT AGENTS AND PHOSPHODIESTERASE INHIBITORS

DRUG CARDS

Make drug cards for the following drugs. (See page 209 for instructions on making drug cards.)

Brand Name	Generic Name	Classification
Detrol LA	tolterodine	Urinary
Avodart	dutasteride	Urinary
Flomax	tamsulosin	Urinary
Vesicare	solifenacin	Urinary
Enablex	darifenacin	Urinary
Uroxatral	alfuzosin	Urinary
Viagra	sildenafil	Hormones & Modifiers, Phosphodiesterase inhibitor
Cialis	tadalafil	Hormones & Modifiers, Phosphodiesterase inhibitor
Levitra	vardenafil	Hormones & Modifiers, Phosphodiesterase inhibitor

LAB ACTIVITIES

1. Visit the MedlinePlus website of the National Library of Medicine, http://www.nlm.nih.gov/medlineplus/.

   a. Search MedlinePlus for Flomax® and select the druginfo result for tamsulosin (http://www.nlm.nih.gov/medlineplus/druginfo/meds/a698012.html) to answer the following questions:

   1. Why is this medication prescribed?
   2. What storage conditions are needed for this medicine?
   3. What are other brand names for tamsulosin products?
   4. Read the pronunciation of tamsulosin at the top of the page and correctly pronounce tamsulosin.

b. Search MedlinePlus for Vesicare® and select the druginfo result for solifenacin (http://www.nlm.nih.gov/medlineplus/druginfo/meds/a605019.html) to answer the following questions:

1. Why is this medication prescribed?
2. What storage conditions are needed for this medicine?
3. What is a special dietary instruction for this product?
4. What are the special precautions for solifenacin products?
5. Read the pronunciation of solifenacin at the top of the page and correctly pronounce solifenacin.

c. Search MedlinePlus for Cialis® and select the druginfo result for tadalafil (http://www.nlm.nih.gov/medlineplus/druginfo/meds/a604008.html) to answer the following questions:

1. Why is this medication prescribed?
2. What storage conditions are needed for this medicine?
3. What is a special dietary instruction for this product?
4. What are the special precautions for tadalafil products?
5. Read the pronunciation of tadalafil at the top of the page and correctly pronounce tadalafil.

2. Prepare a label for each of the following prescriptions using an index card formatted like the example shown on page 209, or prepare a computer-generated label if a computer pharmacy software system is available. Assume you have checked the patient's profile and found that the patient is not taking any other medications, has no medication allergies, and pays cash for the prescriptions.

a. A prescription was written by Dr. Alice Chan for Dustin George (date of birth 7/14/1933) for Flomax® 0.4 mg Disp. #30 g, Sig: i cap daily, 6 refills.

b. A prescription was written by Dr. Alice Chan for Jay Day (date of birth 2/24/1948) for Vesicare® 10 mg Disp. #30 Sig: i daily. 1 refill.

c. A prescription was written by Dr. Alice Chan for Troy Schmidt (date of birth 9/4/1952) for Cialis® 2.5 mg Disp. #15, Sig: i prn prior to sexual activity. 6 refills.

Pharmacy Technician Certification Exam (PTCE)

Practice Exam

The following multiple choice questions are in the *choose the best answer format* of the national Pharmacy Technician Certification Examination (PTCE) offered by the Pharmacy Technician Certification Board (PTCB). There are four possible answers with only one answer being the most correct. Many of these questions can be answered through a careful review of this workbook. However, others require knowledge gained from practice as a technician. Answers for all questions can be found on page 278.

Since the time limit for taking the National Exam is two hours, you may want to test your ability to answer the questions under a time limit, or you may simply wish to time yourself to see how long it takes you. There are 90 questions here, the same number as on the exam. If you wish to have a similar experience, you can allow yourself two hours to answer these questions.

For more information on the Pharmacy Technician Certification Exam, see the preface of this Workbook.

PTCE Practice Exam

Answers are on page 278

1. The pharmacist has asked you to obtain a MedWatch Form 3500 so s/he can report
   a. an adverse event regarding a veterinary product.
   b. an adverse event regarding a drug.
   c. an adverse event regarding a vaccine.
   d. an adverse event regarding a prescribing error.

2. Medications for ophthalmic administration are usually available in
   a. sterile hypotonic drops or sterile ointment.
   b. sterile hypertonic drops or sterile ointment.
   c. hypotonic solution or hypotonic suspension.
   d. sterile isotonic drops or sterile ointment.

3. Risk Evaluation and Mitigation Strategies (REMS) may require a/an
   a. Medication Guide (MedGuide).
   b. PDR.
   c. primary literature.
   d. abstracting service.

4. You receive a prescription for Mary Jones and note there are two patients with that name in your computer system. To prevent errors, the best course of action is
   a. process the prescription for the patient who gets the most prescriptions filled at your pharmacy.
   b. ask a colleague which Mary Jones you should choose.
   c. verify the birthdate of the patient.
   d. process the prescription for the patient with the fewest allergies.

5. An excellent resource for a list of commonly confused drug names is
   a. ISMP www.ismp.org.
   b. APhA www.pharmacist.com.
   c. USP www.usp.org.
   d. NABP www.nabpnet.org.

6. Biological safety cabinets protect personnel and the environment from
   a. light.
   b. temperature.
   c. contamination.
   d. freezing.

7. _______________ is a pharmacy technician responsibility.
   a. Checking that the patient knows how to take the medication
   b. Checking that the patient understands the expected benefits of taking the medication
   c. Counseling patients
   d. Quickly locating the correct medication for dispensing

8. The form number for ordering Schedule II drugs is
   a. DEA Form 121.
   b. DEA Form 200.
   c. DEA Form 222.
   d. DEA Form 240.

9. Patient package inserts (PPIs) for oral contraceptive refills are required to be provided to patients
   a. only for new prescriptions.
   b. for new prescription and refills if 30 days have lapsed since the patient received a PPI.
   c. every six months.
   d. once a year.

10. How much cough syrup will a patient take in 24 hours if the dose is two teaspoonsful every six hours?
    a. 20 mL
    b. 30 mL
    c. 40 mL
    d. 80 mL

PTCE PRACTICE EXAM

11. The smallest gelatin capsule used for extemporaneous compounding is size
    a. 10.
    b. 8.
    c. 5.
    d. 000.

12. Coring can occur when
    a. the needle is longer than the ampule.
    b. the patient has an allergy to latex.
    c. the needle is not correctly removed from the vial.
    d. the needle is not correctly inserted into the vial.

13. Hydrochlorothiazide is used as
    a. an analgesic.
    b. an anti-inflammatory agent.
    c. a sedative.
    d. a diuretic.

14. The infusion rate of an IV is over 12 hours. The total exact volume is 800 mL What would be the infusion rate in mLs per minute?
    a. 0.56 mL/minute
    b. 1.11 mL/minute
    c. 2.7 mL/minute
    d. none of the above

15. You have a 70% solution of dextrose. How many grams of dextrose is in 200 mL of this solution?
    a. 700 grams
    b. 460 grams
    c. 140 grams
    d. 120 grams

16. The standard pediatric dose for cefazolin is 20 mg/kg/day. The order is written for 150 mg TID. The infant weighs 8 lb. This dose is
    a. too high.
    b. too low.
    c. within guidelines.
    d. none of the above.

17. Copy 3 of the DEA Form 222 is kept by the
    a. wholesaler.
    b DEA.
    c. state board.
    d. pharmacy.

18. An IV order calls for the addition of 45 mEq of $CaCO_3$ (calcium carbonate). You have a 25 mL vial of $CaCO_3$ 4.4mEq/mL. How many mL of this concentrate do you need to add to this IV?
    a. 5.6 mL
    b. 8.4 mL
    c. 10.2 mL
    d. 12.8 mL

19. Which of the following is true about high-alert medications?
    a. High-alert medications are medications that are known to cause significant harm to the patient if an error is made.
    b. High-alert medications are medications that are expensive.
    c. High-alert medications are medications that only come in generic.
    d. High-alert medications are always multisource.

20. Which of the following is true about medication guides?
    a. Medication guides are part of the FDA-approved labeling.
    b. Manufacturers are required to provide medication guides for all FDA-approved drugs.
    c. Medication guides are written by the dispensing pharmacist.
    d. Medication guides are written by the technician who fills the prescription.

PTCE PRACTICE EXAM

21. A pharmacy wants to markup a product by 30 percent. How much would an item cost with this markup, if its original cost was $4.50?
    a. $5.85
    b. $6.23
    c. $6.40
    d. $7.10

22. Federal law requires an exact inventory must be kept for
    a. phenobarbital.
    b. alprazolam.
    c. diazepam.
    d. hydrocodone.

23. Safety Data Sheets (SDS)
    a. provide protocols for fire hazards in the pharmacy setting.
    b. provide safety codes by OSHA in the storage of inventory.
    c. provide information concerning hazardous substances.
    d. none of the above

24. How much diluent do you need to add to 4 gm of powder to get a concentration of 500 mg/mL?
    a. 0.8 mL
    b. 8 mL
    c. 1.0 mL
    d. 10 mL

25. Bedside Medication Verification (BMV) in hospitals utilizes
    a. nurse's bar-coded name badge, patient's bar-coded arm band, and medication bar code.
    b. patient's bar-coded arm band and medication bar code only.
    c. nurse's bar-coded name badge and medication bar code only.
    d. nurse's bar-coded name badge and patient's bar-coded arm band only.

26. For most medication stock bottles, the bar code includes
    a. the product's NDC code.
    b. the pharmacist's license number.
    c. the technician's registration number.
    d. the DEA number of the pharmacy.

27. The last set of digits of the NDC are indicative of
    a. the manufacturer.
    b. product identification.
    c. package size.
    d. none of the above.

28. The approximate size container for the dispensing of 180 mL of liquid medication would be?
    a. 2 ounces
    b. 4 ounces
    c. 6 ounces
    d. 8 ounces

29. A patient asks whether he/she can take a certain medication with another one? As a pharmacy technician what should you do?
    a. Inform the patient that you see no problem.
    b. Provide the patient with a drug insert.
    c. Request the patient see the pharmacist for a consult.
    d. Try to sell the patient some Tylenol®.

30. The doctor writes: ii gtts as bid. What does this mean?
    a. two drops in the left eye twice a day
    b. two drops in the right eye twice a day
    c. two drops in the left ear twice daily.
    d. two drops in the right ear twice a day

31. Investigational drugs are regulated by the
    a. FDA.
    b. CMS.
    c. DEA.
    d. Board of Pharmacy.

PTCE PRACTICE EXAM

32. Which of the following is used for the FDA's list of approved drug products?
    a. *Merck Index*
    b. *RED BOOK*
    c. "Orange Book"
    d. *Martindale: The Complete Drug Reference*

33. To help prevent errors, which drugs should not be placed next to each other?
    a. Hydroxyzine and hydralazine
    b. Bactroban® ointment and hydrocortisone cream
    c. Timolol ophthalmic and Alphagan® ophthalmic
    d. Crestor® and Dilantin®

34. Companies that specialize in returns of expired and discontinued drugs to the manufacturer are known as
    a. reverse distributors.
    b. pharmacy benefit managers.
    c. mail order pharmacy.
    d. mass merchandiser pharmacy.

35. Of the following, which one deals with the issue of safety caps on prescription bottles?
    a. The Controlled Substance Act
    b. The Poison Prevention Act
    c. Hazardous Substance Act
    d. Federal Food and Cosmetic Act

36. Most drugs are kept at room temperature between
    a. 33°–45°F.
    b. 33°–45°C.
    c. 59°–86°C.
    d. 59°–86°F.

37. The appearance of crystals in mannitol injection would indicate that the product
    a. was exposed to cold.
    b. has settled during shipment.
    c. contains impurities and should be returned.
    d. was formulated using sterile saline.

38. The rules for coordination of benefits ensure that the benefit coverage for a claim does not exceed _______ of the total cost.
    a. 25%
    b. 33%
    c. 50%
    d. 100%

39. Dextrose 25% 1,000 mL is ordered. You have only dextrose 70% solution available. How much of the dextrose 70% solution and sterile water will you use to fill this order?
    a. 250 mL dextrose 70% and 750 mL sterile water
    b. 357 mL dextrose 70% and 643 mL sterile water
    c. 424 mL dextrose 70% and 576 mL sterile water
    d. None of the above

40. The Occupational, Safety & Health Administration (OSHA) requires pharmacies to have Safety Data Sheets (SDS) for
    a. all drugs in the pharmacy inventory.
    b. all materials stored in the pharmacy refrigerator.
    c. each hazardous chemical used in the pharmacy.
    d. all controlled substances in the pharmacy inventory.

PTCE PRACTICE EXAM

41. Of the following group names, which one would be used for cough?
    a. Anthelmintics
    b. Antitussives
    c. Antihistamines
    d. Anticholinergics

42. Tobrex® ophthalmic ung refers to
    a. an ointment used for the eye.
    b. a solution used for the eye.
    c. a topical ointment for external use only.
    d. an ointment used for the ear.

43. Suspending or thickening agents are added to suspensions to thicken the suspending medium and the sedimentation rate. Which of the following is not a suspending agent?
    a. Carboxymethylcellulose
    b. Tragacanth
    c. Acacia
    d. Bentonite

44. Oral polio virus vaccine (Poliovax®) should be stored in a temperature not to exceed 46 degrees Fahrenheit. What is this temperature in Centigrade? Use this formula: Centigrade = 5/9 (F° - 32°).
    a. 6°C
    b. 8°C
    c. 10°C
    d. 12°C

45. You receive a prescription for amoxicillin 125 mg TID for 10 days. How many mL of amoxicillin 250 mg/5mL do you need to fill this prescription to last the full 10 days?
    a. 20 mL
    b. 40 mL
    c. 75 mL
    d. 100 mL

46. Medicare Part D covers
    a. prescription drugs.
    b. doctors' services.
    c. inpatient hospital expenses.
    d. hospice expenses.

47. You receive a prescription for sertraline (Zoloft®) qd x 30 days. What is sertraline?
    a. Antihypertensive
    b. Anticonvulsant
    c. Antidepressant
    d. Antianginal

48. All aseptic manipulations in the laminar airflow workstation should be performed at least
    a. four inches within the workstation.
    b. six inches within the workstation.
    c. eight inches within the workstation.
    d. twelve inches within the workstation.

49. Which auxiliary label would be used for a prescription for tetracycline 250 mg capsules?
    a. May Cause Drowsiness
    b. Avoid Aspirin
    c. Avoid Dairy Products and Antacids
    d. Take with Food

50. Generic drugs are usually in tier
    a. 1.
    b. 2.
    c. 3.
    d. 4.

51. Which of the following is a Schedule II Controlled Substance?
    a. Diazepam
    b. Meperidine
    c. Pentazocine
    d. Diphenoxylate, atropine

PTCE PRACTICE EXAM

52. If the manufacturer's expiration date for a drug is 12/17, the drug is considered acceptable to dispense until which date?
    a. 12/01/17
    b. 12/31/17
    c. 11/30/17
    d. 1/01/17

53. HMOs, POSs, and PPOs are examples of
    a. MAC.
    b. co-insurance.
    c. managed care programs.
    d. co-pays.

54. The laminar airflow workstation (LAFW) should be left operating continuously. If it is turned off, it should not be used until it has been running for at least
    a. 10 minutes.
    b. 30 minutes.
    c. 45 minutes.
    d. 60 minutes.

55. Which auxiliary label would you use for this particular sig: ii gtts AU bid?
    a. Take with meals
    b. For the ear
    c. Avoid sunlight
    d. For the eye

56. A dose is written for 10 mg/kg every eight hours for one day. The adult to take this medication weighs 145 pounds. How much drug will be needed to fill this order?
    a. 1,530 mg
    b. 1,688 mg
    c. 1,978 mg
    d. 2,508 mg

57. How much medication would be needed for the following order?

    prednisone 10 mg, one qid x 4 days, one tid x 2 days, one bid x 1 day, then stop
    a. 16
    b. 20
    c. 24
    d. 26

58. Benzethidine is in DEA Schedule I, meaning that benzethidine
    a. has a currently accepted medical use in the United States with severe restrictions.
    b. can only be handled by the pharmacist.
    c. has no currently accepted medical use in the world.
    d. has no currently accepted medical use in the United States.

59. In which controlled substance schedule is Tylenol® No. 2 classified?
    a. Schedule I
    b. Schedule II
    c. Schedule III
    d. Schedule IV

60. Where would a pharmacy technician look for nationally recognized AWPs and NDCs for FDA-approved drugs?
    a. *AHFS*
    b. "Orange Book"
    c. *RED BOOK*
    d. *Remington: The Science & Practice of Pharmacy*

61. The first line of defense against infection/contamination of an IV product is
    a. antibiotics.
    b. antiseptics.
    c. disinfectants.
    d. handwashing.

PTCE PRACTICE EXAM

62. HIPAA requires that
    a. all Medicaid patients are offered counseling by a pharmacist.
    b. all patients receive counseling by a pharmacist.
    c. all patients are offered counseling by a pharmacist.
    d. privacy rules are observed for PHI.

63. Which of the following medications must be administered in a glass IV container?
    a. Aminophylline
    b. Dopamine
    c. Nitroglycerin
    d. Potassium

64. The two parts of the syringe that should not be touched are
    a. the tip and needle.
    b. the collar and barrel.
    c. the tip and plunger.
    d. the collar and plunger.

65. The sale of what medication is restricted by the Combat Methamphetamine Epidemic Act (CMEA)?
    a. Claritin®
    b. Claritin-D 24®
    c. Chlor-Trimeton®
    d. Robitussin DM®

66. The first five digits of the National Drug Code (NDC) number identifies the
    a. product.
    b. manufacturer.
    c. units.
    d. type of packaging.

67. The "Orange Book" provides information about
    a. current pricing.
    b. drug product stability.
    c. generic equivalents.
    d. investigational drugs.

68. The common name for the FDA's Approved Drug Products with Therapeutic Equivalent Evaluations is the
    a. "Green Book."
    b. "Orange Book."
    c. "Red Book."
    d. "Blue Book."

69. What should the last digit be of this DEA number? AB431762 __
    a. 1
    b. 3
    c. 5
    d. 7

70. Aminosyn® is an amino acid often used in TPN orders to provide protein for cellular repair and growth. A physician writes an order for Aminosyn® 2.5% 500 mL. You have only Aminosyn® 8.5% 500 mL. How do you prepare this order using a sterile evacuated container?
    a. Add 320 mL of Aminosyn® 8.5% and qs with sterile water to 500 mL.
    b. Add 147 mL of Aminosyn® 8.5% and qs with sterile water to 500 mL.
    c. Add 124 mL of Aminosyn® 8.5% and qs with sterile water to 500 mL.
    d. Add 74 mL of Aminosyn® 8.5% and qs with sterile water to 500 mL.

71. VIPPS provides accreditation for
    a. technician programs.
    b. Internet pharmacies.
    c. hospital pharmacies.
    d. community pharmacies.

PTCE PRACTICE EXAM

72. Hard copies of order reports
    a. are kept for an established amount of time for business and legal reasons.
    b. are only needed if there is a computer failure.
    c. are only needed in the event of a power failure.
    d. are no longer needed since everything is computerized.

73. The type of formulary that allows the pharmacy to obtain all medications that are prescribed is a(an)
    a. international formulary.
    b. closed formulary.
    c. wholesaler formulary.
    d. open formulary.

74. Zantac®, Tagamet®, and Pepcid® are H2 blockers that are now available over-the-counter (OTC). What are these drugs used for?
    a. As an antihistamine to alleviate runny nose
    b. As a decongestant to unclog nasal passages
    c. To inhibit stomach acid secretion
    d. As an antacid in that it neutralizes stomach acid

75. A ____________________ is an inventory system in which the item is deducted from inventory as it is sold or dispensed.
    a. reorder point system
    b. point-of-sale system
    c. turnover system
    d. automated system

76. Of the following drug recalls, which one is the most important in that all parties involved in the dispensing of a prescription (doctor, pharmacy, and patient) must be notified due to the drugs potential for serious harm?
    a. Class I Recall
    b. Class II Recall
    c. Class III Recall
    d. Class IV Recall

77. The pharmacist should be alerted if a patient is allergic to codeine and prescribed
    a. alprazolam.
    b. Tylenol #3®.
    c. methylphenidate.
    d. phenobarbital.

78. A prescription for amoxicillin 250 mg #30 has a usual and customary price of $8.49. The acquisition cost of amoxicillin 250 mg #30 is $2.02. What is the gross profit?
    a. $2.02
    b. $6.47
    c. 50%
    d. 1/3

79. A senior citizen is paying for a prescription for penicillin VK 250 mg #30. The usual and customary price is $8.49. However this patient qualifies for a 10% discount. How much will the patient pay?
    a. $8.49
    b. $6.99
    c. $8.39
    d. $7.64

80. A medication used to reduce a fever is called an
    a. antipyretic.
    b. antitussive.
    c. antiemetic.
    d. anthelmintic.

PTCE PRACTICE EXAM

81. A prescription is written for dicyclomine syrup 10 mg/5 mL 1 teaspoonful h.s. + 1 refill. The insurance plan has a 34-day supply limitation. How many mL can be dispensed using the insurance plan guidelines?
   a. 120 mL
   b. 170 mL
   c. 240 mL
   d. 360 mL

82. A prescription is written for Humulin® N U-100 insulin 10 mL, 40 units daily. What is the days supply?
   a. 25
   b. 34
   c. 21
   d. 28

83. A prescription is written for Tetracycline HCl suspension 125 mg/5 mL compounded from capsules and a mixture of Ora-Plus® 50% and Ora-Sweet® 50%. How many capsules of Tetracycline 250 mg are needed to prepare 50 mL of this suspension?
   a. 5
   b. 10
   c. 15
   d. 20

84. How many units of insulin does a 1/3 mL insulin syringe hold?
   a. 0.3
   b. 25
   c. 30
   d. 50

85. Nurses track medication administration on a(an)
   a. STAT.
   b. MAR.
   c. IVP.
   d. IVPB.

86. When entering a new prescription into the pharmacy computer, the technician must enter the following information:
   a. The prescription number
   b. The directions for use
   c. The DEA number of the wholesaler
   d. The date the bottle was opened

87. Alprozolam is a
   a. narcotic.
   b. barbiturate.
   c. benzodiazepine.
   d. stimulant.

88. A prescription for Duragesic® patches should be filed under which DEA schedule?
   a. Schedule I
   b. Schedule II
   c. Schedule III
   d. Schedule IV

89. Which of the following medications is an antidiarrheal?
   a. Propranolol
   b. Famotidine
   c. Methylphenidate
   d. Loperamide

90. If a medication is to be taken a.c., it should be taken
   a. in the morning.
   b. around the clock.
   c. after meals.
   d. before meals.

THE EXAM FOR THE CERTIFICATION OF PHARMACY TECHNICIANS (EXCPT)

PRACTICE EXAM

The following multiple choice questions are in the *choose the best answer* format of the national Exam for the Certification of Pharmacy Technicians (ExCPT) offered by the Institute for the Certification of Pharmacy Technicians (ICPT). There are four possible answers with only one answer being the most correct. Many of these questions can be answered through a careful review of this workbook. However, others require knowledge gained from practice as a technician. Answers for all questions can be found on page 278.

Since the time limit for taking the ExCPT is two hours and 10 minutes, you may want to test your ability to answer the questions under a time limit, or you may simple wish to time yourself to see how long it takes you. There are 120 questions here, the same number as on the ExCPT exam.

For more information on the Exam for the Certification of Pharmacy Technicians, see the preface of this Workbook.

EXCPT PRACTICE EXAM

Answers are on page 278.

1. Which of the following may be performed only by a pharmacist?
   a. Accepting a call from a wholesaler about an order
   b. Accepting a return call from a prescriber's office clarifying a prescription
   c. Calling a prescriber on behalf of a patient to request refills
   d. Calling an insurance company to verify a patient's eligibility

2. All of the following duties may be performed by a pharmacy technician EXCEPT
   a. requesting PHI from a patient such as date of birth, address, allergy, and insurance information.
   b. selecting an OTC product for a patient.
   c. inputting and updating patient information in the computer.
   d. placing the medication in a vial and attaching the prescription label to it.

3. A patient brings in two new prescriptions from two different doctors. One prescription is for ibuprofen 800 mg, 1 tablet po t.i.d. and another prescription is for Anaprox DS®, 1 tablet b.i.d. What should the technician do?
   a. Tell the patient to decide which prescription s/he wants to fill
   b. Tell the patient that s/he must get generic for both prescriptions
   c. Alert the pharmacist of a possible therapeutic duplication
   d. Fill both prescriptions as written

4. A list of the goods or items a business will use in its normal operation is called a (an)
   a. open formulary.
   b. closed formulary.
   c. inventory.
   d. protocol.

5. Checking order reports to ensure the order contains no errors is done
   a. by computer.
   b. manually.
   c. by the wholesaler.
   d. by the corporate office.

6. In receiving an order, it is important to be alert for drugs that have been incorrectly picked, received damaged, are outdated, or missing, so orders are reconciled
   a. item by item.
   b. once per month.
   c. once per year.
   d every two years.

7. A small volume intravenous bag specifically used to deliver medication is called an
   a. LVP.
   b. IVPB.
   c. vial.
   d. ampule.

8. According to federal law, the prescriber must provide his/her DEA number for which of the following prescriptions?
   a. Ultram®
   b. Xanax®
   c. Wellbutrin®
   d. Prozac®

9. Isotretinoin is associated with
   a. iPLEDGE.
   b. RevAssist.
   c. STEPS.
   d. TOUCH.

10. SDS sheets provide
    a. protocols for pharmacy evacuations.
    b. safety codes from OSHA.
    c. ordering numbers from the wholesaler.
    d. information about hazardous substances.

ExCPT Practice Exam

11. An antibiotic is prescribed 15 mg/kg twice a day by IV. What is the daily dose in mg for a child weighing 66 pounds?
    a. 225 mg
    b. 450 mg
    c. 675 mg
    d. 900 mg

12. Outdated drugs can be
    a. sent to reverse distributors.
    b. shipped to the DEA.
    c. shipped to the state board of pharmacy.
    d. stocked with medications that are to be dispensed.

13. An example of a major drug-drug interaction is
    a. warfarin-aspirin.
    b. hydrocodone-codeine.
    c. guaifenesin-pseudoephedrine.
    d. hydrochlorothiazide-triamterene.

14. The patient's ________ is necessary for billing a prescription.
    a. birth date
    b. medication history
    c. allergy information
    d. disease information

15. As you are putting away an order from a wholesaler, you notice one bottle is missing a label. The pharmacist tells you that the medication cannot be dispensed and must be returned to the wholesaler or destroyed. The drug cannot be dispensed because it is
    a. adulterated.
    b. misbranded.
    c. exempt.
    d. recycled.

16. Companies that administer drug benefit programs are called
    a. HMOs.
    b. PBMs.
    c. MACs.
    d. PPOs.

17. Copy 2 of DEA Form 222 is
    a. retained by the pharmacy.
    b. retained by the wholesaler.
    c. retained by the manufacturer.
    d. forwarded to the DEA.

18. Proper disposal or destruction of non-returnable medications includes
    a placement in regular trash.
    b. flushing down a sink or water system.
    c. using a company that meets EPA regulations.
    d. selling the medications at a discount.

19. Each tablet of Claritin D® 12 hour contains 120 mg of pseudoephedrine sulfate. What is the maximum number of tablets that could be sold in a single transaction if the maximum amount per transaction of pseudoephedrine is 3.6 g?
    a. 15
    b. 20
    c. 30
    d. 36

20. Exempt narcotics are in DEA schedule
    a. II.
    b. III.
    c. IV.
    d. V.

21. Exempt narcotics are regulated for how many dosage units can be sold without a prescription in a ______ hour period.
    a. 24
    b. 48
    c. 72
    d. 168

ExCPT Practice Exam

22. Federal law requires pharmacies to keep ________ of the DEA Form 222.
    a. Copy 1
    b. Copy 2
    c. Copy 3
    d. Copy 4

23. For aseptic technique, all work should be performed at least _____ inches inside the laminar flow hood.
    a. 2
    b. 4
    c. 6
    d. 8

24. All of the following are used to store medications EXCEPT
    a. automated dispensing machines.
    b. carousels.
    c. Pyxis machines.
    d. shoe boxes.

25. Counseling as required by OBRA is provided by
    a. physicians.
    b. physician assistants.
    c. certified pharmacy technicians.
    d. pharmacists.

26. All of the following pharmacy personnel must be formally trained on HIPAA EXCEPT
    a. pharmacy technicians.
    b. pharmacists.
    c. pharmacy clerks.
    d. none of the above.

27. For the following prescription, how much drug is taken each day?

    Amoxicillin 250 mg/5 mL 150 mL, Sig: 2 teaspoons t.i.d.

    a. 15 mL
    b. 30 mL
    c. 45 mL
    d. 50 mL

28. HIPAA is a federal law that protects a patient's PHI when it is
    a. written.
    b. electronically transferred.
    c. spoken.
    d. all of the above.

29. The directions on a prescription read ii gtt q8h ou. The directions on the label should read:
    a. Instill one drop in the right ear every eight (8) hours.
    b. Inhale two puffs orally every eight (8) hours.
    c. Instill two drops in each eye every eight (8) hours.
    d. Take two drops by mouth every eight (8) hours.

30. How many cases of 8 dram vials should be ordered to last two weeks if there are 500 vials per case and the pharmacy uses 120 vials per day on average?
    a. 2
    b. 4
    c. 6
    d. 8

31. The common abbreviation ss means
    a. safety service
    b. left ear.
    c. left eye.
    d. one-half.

32. Which abbreviation is on The Joint Commission's "Do Not Use List"?
    a. qd
    b. prn
    c. ac
    d. mg

EXCPT PRACTICE EXAM

33. How many grams of sucrose are needed to prepare 4 ounces of a 20% solution?
    a. 6
    b. 12
    c. 24
    d. 36

34. How many milliliters of magnesium sulfate 4 mEq/mL are needed to add 10 mEq of magnesium sulfate to an IV bag?
    a. 2
    b. 2.5
    c. 4
    d. 10

35. Aseptic techniques are methods used to maintain
    a. pH.
    b. temperature.
    c. osmolarity.
    d. sterility.

36. How much hydrocortisone powder is needed to compound 60 g of hydrocortisone cream 2%?
    a. 1.2 mg
    b. 1,200 mg
    c. 0.06 g
    d. 6 g

37. Good interpersonal skills include
    a. making eye contact.
    b. calling the patient by name.
    c. listening carefully.
    d. all of the above.

38. How much should be withdrawn from a vial if the strength of the medication is 1 mg/mL and a dose of 5,000 micrograms is needed?
    a. 0.25 mL
    b. 0.5 mL
    c. 1 mL
    d. 5 mL

39. How should a prescription for BenzaClin® be stored after it has been mixed?
    a. Refrigerator
    b. Freezer
    c. Room temperature
    d. In a warm and moist environment as defined by USP

40. Which dispensing code should be used for a prescription written for Cymbalta 20 mg if DAW was not written on the prescription?
    a. 0
    b. 1
    c. 2
    d. 3

41. If a medication is to be taken p.c., when should it be taken?
    a. After meals
    b. In the evening
    c. Before meals
    d. In the morning

42. Which dispensing code should be used for a prescription written for Norvasc 5 mg if the patient requested brand name and DAW was not written on the prescription?
    a. 0
    b. 1
    c. 2
    d. 3

43. If the dose of liquid amoxicillin for a child is one and one-half teaspoonsful three times a day for 10 days, what is the volume needed to fill the prescription?
    a. 100 mL
    b. 150 mL
    c. 200 mL
    d. 225 mL

EXCPT PRACTICE EXAM

44. Information about generic equivalents can be found in the
    a. *RED BOOK.*
    b. "Orange Book."
    c. blue book.
    d. green book.

45. Tall man letters are used
    a. to distinguish look-alike, sound-alike drug names.
    b. to promote brand products.
    c. for growth hormones.
    d. for fast movers.

46. MedWatch Form 3500 is for reporting adverse reactions to
    a. vaccines.
    b. drugs.
    c. veterinary products.
    d. produce.

47. Normal saline solution is
    a. hypotonic.
    b. isotonic.
    c. hypertonic.
    d. hyperosmotic.

48. OSHA required notices for hazardous substances that provide hazard, handling, clean-up, and first aid information are called
    a. MAC.
    b. SDS.
    c. MEC.
    d. HCFA-1500.

49. When a technician receives a rejected claim "NDC Not Covered," this probably means
    a. the insurance plan has a closed formulary.
    b. the insurance plan has an open formulary.
    c. the birth date submitted does not match the birth date on file.
    d. the patient has single coverage.

50. The age required to purchase a CV exempt narcotic without a prescription is
    a. 16 years and older.
    b. 18 years and older.
    c. 19 years and older.
    d. 21 years and older.

51. Premarin® cream has been prescribed 0.5 g pv twice a week. The medication should be administered
    a. vaginally.
    b. rectally.
    c. on the stomach.
    d. on the forearms.

52. Lidoderm® is an example of an
    a. analgesic.
    b. antianxiety.
    c. anesthetic.
    d. antidepressant.

53. What is the % concentration if 134 g of sugar is in 200 mL of aqueous solution?
    a. 33%
    b. 50%
    c. 67%
    d. 75%

54. Sublingual nitroglycerin is used for
    a. hyperlipidemia.
    b. angina.
    c. hypertension.
    d. coagulation.

55. The term for a drug that reduces fever is
    a. analgesic.
    b. antipyretic.
    c. anti-inflammatory.
    d. antidepressant.

ExCPT Practice Exam

56. When a technician receives a rejected claim "Invalid Person Code," this probably means
    a. the patient is on Medicare.
    b the patient has a mail order program.
    c. the person code entered does not match the birth date and/or sex in the insurer's computer.
    d. the patient is on Medicaid.

57. When a technician receives a rejected claim "Unable to Connect," this probably means
    a. the insurer has an incorrect birth date for the patient.
    b. the patient's coverage has expired.
    c. the connection with the insurer's computer is temporarily unavailable due to computer problems.
    d. the insurer has a closed formulary.

58. The class of drugs to dissolve blood clots is
    a. thrombolytics.
    b. vasopressors.
    c. antianginals.
    d. antihypertensives

59. The CMEA sets ______ and ______ limits on the over-the-counter sale of pseudoephedrine.
    a. daily, weekly
    b. weekly, monthly
    c. daily, monthly
    d. monthly, yearly

60. The commission that surveys and accredits health-care organizations is the
    a. DPH.
    b. FDA.
    c. ASHP.
    d. TJC.

61. Nitroglycerin sublingual tablets should be dispensed
    a. in a glass container.
    b. in a plastic container.
    c. with a cotton plug.
    d. with a medicine dropper.

62. The Controlled Substances Act is enforced by the
    a. FDA.
    b. CPSC.
    c. DEA.
    d. Bureau of Alcohol, Tobacco and Firearms.

63. The directions state 2 gtt a.u. q12h. How is the drug to be administered?
    a. Both ears
    b. Right eye
    c. By mouth
    d. Topically

64. The CPT codes for billing Medication Therapy Management services provided by pharmacists are
    a. MAC.
    b. PPO.
    c. ICD-9.
    d. 99605, 99606, and 99607.

65. Which of the following types of information is generally not required for online claim processing?
    a. Birth date
    b. Sex
    c. Group number
    d. Height

66. The following drugs are antidepressants EXCEPT
    a. citalopram.
    b. trazodone.
    c. venlafaxine.
    d. diazepam.

ExCPT Practice Exam

67. The following drugs are considered antihyperlipidemic drugs EXCEPT
    a. Pravachol®.
    b. Tricor®.
    c. Vytorin®.
    d. Aldactone®.

68. The following medications should be refrigerated EXCEPT
    a. Biaxin® suspension.
    b. Augmentin® suspension.
    c. Miacalcin®.
    d. Thyrolar®.

69. The generic name for Neurontin® is
    a. gabapentin.
    b. etodolac.
    c. nabumetone.
    d. finasteride.

70. The government agency that regulates investigational medications is the
    a. FDA.
    b. DEA.
    c. CPSC.
    d. USP.

71. An example of a robotic dispensing machine is
    a Pyxis.
    b. Parata Max.
    c. Kirby Lester.
    d. a carousel.

72. The middle set of digits of the NDC represents
    a. the manufacturer.
    b. product identification.
    c. package size.
    d. country of origin.

73. Patient Package Inserts (PPIs) are required to be dispensed with all new prescriptions and at least every 30 days with refill prescriptions for
    a. oral contraceptives.
    b. antibiotics.
    c. cough syrups.
    d. NSAIDs.

74. The pharmacist should be alerted if a patient is allergic to sulfa and prescribed
    a. Ceftin®.
    b. Keflex®.
    c. Erythrocin®.
    d. Bactrim®.

75. The *RED BOOK* provides information about
    a. pricing.
    b. stability.
    c. solubility.
    d. generic substitution.

76. Prescriptions for controlled substances from Schedules II, III, and IV must contain the following warning
    a. Take with food.
    b. May cause drowsiness.
    c. Caution: Federal law prohibits the transfer of this drug to any person other than the patient for whom it was prescribed.
    d. Take on an empty stomach.

77. The temperature of a refrigerator in a pharmacy should generally be
    a. 33–45°F.
    b. 40–42°F.
    c. 43–55°F.
    d. 50–52°F.

ExCPT Practice Exam

78. The type of formulary that allows the pharmacy to obtain only drugs listed on the formulary with no exceptions is a/an
    a. open formulary.
    b. closed formulary.
    c. third-party formulary.
    d. P&T list.

79. The type of medication order for medication to be administered only on an as-needed basis is called a
    a. standing order.
    b. PRN order.
    c. STAT order.
    d. standard order.

80. The unique identifying number of a drug is called a/an
    a. NDC.
    b. PHI.
    c. NPI.
    d. DEA.

81. Theft or loss of controlled substances must be recorded on DEA Form
    a. 41.
    b. 106.
    c. 222.
    d. 224.

82. When compounds are prepared from solids dissolved in water, the beyond-use date should not be later than _______ when stored in the refrigerator.
    a. 5 days
    b. 7 days
    c. 10 days
    d. 14 days

83. The type of recall for drug products most likely to cause serious adverse effects or death is
    a. Class I.
    b. Class II.
    c. Class III.
    d. Class V.

84. Under HIPAA, who *cannot* have access to information about a patient's prescription?
    a. The patient
    b. The pharmacist & support staff
    c. The physician/prescriber and support staff
    d. The pharmaceutical companies or their representatives

85. Under the Combat Methamphetamine Epidemic Act, what is the maximum amount of pseudoephedrine that can be sold per month to an individual?
    a. 3.6 g
    b. 9 g
    c. 10 g
    d. 20 g

86. Warfarin is a common
    a. thrombolytic.
    b. anticoagulant.
    c. vasopressor.
    d. vasodilator.

87. What is the cost for 60 tablets of a drug if the cost for 100 tablets is $75.00?
    a. $39.30
    b. $43.66
    c. $45
    d. $48.46

ExCPT Practice Exam

88. Which type of mortar and pestle is recommended for mixing liquids and semisolids?
    a. Wedgewood
    b. Porcelain
    c. Glass
    d. Earthenware

89. When compounding parenterals, multi-dose vials
    a. can be reused within 24 hours if they are refrigerated.
    b. can be used within 48 hours if they are refrigerated.
    c. do not contain preservatives.
    d. contain preservatives.

90. What size insulin syringe is needed to deliver a dose of 60 units?
    a. 0.3 mL
    b. 0.5 mL
    c. 1 mL
    d. 2 mL

91. Irrigation solutions are administered
    a. through a filter needle.
    b. through a special administration set.
    c. orally.
    d. by pouring from a bottle.

92. Which class of drugs is used to treat elevated blood lipids?
    a. Keratolytics
    b. Anthelmintics
    c. Antihyperlipidemics
    d. Antihistamines

93. Which dispensing code should be entered if a physician has ordered Coumadin DAW?
    a. 0
    b. 1
    c. 2
    d. 3

94. Which drug does not require child resistant packaging according to the Poison Prevention Packaging Act?
    a. Nitroglycerin sublingual tablets
    b. Isosorbide dinitrate oral tablets
    c. Nitroglycerin sustained release capsules
    d. Isosorbide mononitrate oral tablets

95. If a laminar airflow workstation is turned off between aseptic processing sessions, how long should it run before it is used again?
    a. Does not matter.
    b. At least 15 minutes.
    c. At least 30 minutes.
    d. It should never be turned off.

96. Which drug is an NSAID?
    a. Bumetanide
    b. Nabumetone
    c. Doxazosin
    d. Meclizine

97. Which information is not required by federal law on a prescription label?
    a. Date
    b. Patient's name
    c. Prescriber's name
    d. National drug code

98. Which medication is ordered using DEA Form 222?
    a. Alprazolam
    b. Phenobarbital
    c. Diazepam
    d. Morphine sulfate

99. Which of the following drugs is a controlled substance?
    a. Phenobarbital
    b. Carbamazepine
    c. Imipramine
    d. Furosemide

ExCPT PRACTICE EXAM

100. Which of the following drugs is considered a diuretic?
   a. Furosemide
   b. Lovastatin
   c. Hydroxyzine
   d. Hydralazine

101. Which of the following drugs is considered an analgesic?
   a. Tramadol
   b. Phenelzine
   c. Pidocaine
   d. Diazepam

102. Which of the following drugs is NOT an antihistamine?
   a. Diphenhydramine
   b. Paroxetine
   c. Desloratadine
   d. Cetirizine

103. Which of the following medications does not require ordering with DEA Form 222?
   a. Numbutal®
   b. Percocet®
   c. Codeine
   d. Ultram®

104. A heparin lock is used with a/an
   a. short piece of tubing.
   b LVP solution.
   c. TPN solution.
   d. peristaltic pump.

105. While preparing a prescription, a bottle of finasteride is spilled on the floor and some tablets are broken. Who should stay away from the spill area?
   a. Males with BPH
   b. Females in menopause
   c. Females who are or may become pregnant
   d. Males with hypertension

106. Who is authorized to sign a DEA Form 222 order form?
   a. Pharmacy technicians
   b. Licensed pharmacists
   c. Individuals with power of attorney
   d. Pharmacy managers

107. You are preparing a prescription for Patanol®. Where in the pharmacy should you look for the drug?
   a. Ophthalmics and otics
   b. Tablets and capsules
   c. Creams and ointments
   d. Refrigerator

108. You are putting away an order from a wholesaler and notice the packages in one order crate are floating in water. The pharmacist tells you that the medications in that crate cannot be dispensed and must be destroyed or returned to the wholesaler. These drugs cannot be dispensed because they are
   a. adulterated.
   b. misbranded.
   c. exempt.
   d. recycled.

109. Which route of administration is least likely to give a systemic effect?
   a. Oral
   b. Sublingual
   c. Rectal
   d. Intradermal

110. Zostavax® should be stored at 5°F or colder. What temperature is this in degrees C?
   a. 0°
   b. -5°
   c. -10°
   d. -15°

ExCPT Practice Exam

111. Demerol® is an analgesic used to treat severe pain. What is the generic name?
    a. Ibuprofen
    b. Meperidine
    c. Naproxen
    d. Morphine

112. Antipyretic drugs are used to reduce fever. An example of an antipyretic drug is
    a. morphine.
    b. ibuprofen.
    c. meperidine.
    d. codeine.

113. A patient has been prescribed Bactrim®. What is an important caution for this prescription?
    a. It is only available as an injection.
    b. It should be kept in the refrigerator.
    c. It is associated with extreme drowsiness.
    d. It is important to drink plenty of water.

114. What is the generic name for Tenormin®?
    a. Metoprolol
    b. Nadalol
    c. Propranolol
    d. Atenolol

115. What is the generic name for Calan®?
    a. Nifedipine
    b. Diltiazem
    c. Verapamil
    d. Amlodipine

116. An example of a diuretic that can cause hypokalemia is
    a. spironolactone.
    b. triamterene.
    c. hydrochlorothiazide.
    d. amlodipine.

117. The generic name for Vasotec® is
    a. enalapril.
    b. simvastatin.
    c. ramipril.
    d. pravastatin

118. Lactated Ringer's solution is
    a. acidic.
    b. hypertonic.
    c. hypotonic.
    d. isotonic

119. A parenteral nutrition prescription calls for 0.05 mL/kg trace elements. How many mL are needed for a 55 kg patient?
    a. 1.25 mL
    b. 1.5 mL
    c. 2.25 mL
    d. 2.75 mL

120. A compounded prescription requires 5 g of tetracycline. How many 500 mg capsules are needed?
    a. 10
    b. 20
    c. 50
    d. 100

CALCULATIONS PRACTICE EXAM

The following 50 multiple choice problems provide you with extra practice with pharmacy calculations before taking a national certification exam. There are four possible answers for each problem, with only one answer being the most correct. Similar to the questions in the certification exams, many of the problems in the Calculations Practice Exam can be solved using techniques that are included in this workbook; however, others require knowledge gained from practice as a technician. Answers for all questions can be found on page 278.

Pharmacy calculations must be carefully done with 100% accuracy. Therefore, you should practice doing calculations using a systematic approach for problem solving and always double-check your work.

CALCULATIONS PRACTICE EXAM

Answers are on page 278.

1. You are filling a prescription that reads: Amoxicillin 125 mg/5 mL, Sig 1 tsp t.i.d. Dispense 150 mL. How many milliliters should the patient take each day?
   a. 5 mL
   b. 10 mL
   c. 15 mL
   d. 20 mL

2. You are filling a prescription that reads: EES® 200, Sig 1 tsp t.i.d. Dispense 150 mL. How many days should this prescription last?
   a. 5 days
   b. 7 days
   c. 10 days
   d. 14 days

3. A compounded prescription calls for 600 g of white petrolatum. How many 1 lb. jars should you obtain so there is enough to prepare this prescription?
   a. 1 jar
   b. 2 jars
   c. 3 jars
   d. 4 jars

4. Each tablet of Tylenol® #3 has 30 mg of codeine. How many grains of codeine are in each tablet of Tylenol® #3?
   a. 1/2 grain
   b. 1 grain
   c. 2 grains
   d. 3 grains

5. One form of influenza vaccine, the Live Attenuated Intranasal Vaccine, must be stored at -15°C. What is this temperature in degrees F?
   a. 5°F
   b. -5°F
   c. 27°F
   d. -27°F

6. The lead pharmacy technician must reorder billing forms when 80% of the case has been used. A full case of billing forms contains 20 packages of forms. How many packages of forms should be remaining in the case when it's time to reorder?
   a. 16 packages
   b. 12 packages
   c. 8 packages
   d. 4 packages

7. In reconstituting a liquid antibiotic, 90 mL of distilled water should be used. If 1/3 of the water should be added first to moisten the powder, how much water should be added to moisten the powder?
   a. 45 mL
   b. 60 mL
   c. 15 mL
   d. 30 mL

8. How many capsules of clindamycin hydrochloride should be used to prepare 30 mL of the following preparation, if each capsule contains 150 mg of clindamycin hydrochloride?

clindamycin hydrochloride	600 mg
70% isopropyl alcohol	qs ad 60 mL

   a. 4 capsules
   b. 2 capsules
   c. 10 capsules
   d. 20 capsules

9. A prescription is written for ibuprofen 10% cream. How much ibuprofen is needed to make 20 g of the cream?
   a. 400 mg
   b. 200 mg
   c. 2 g
   d. 4 g

CALCULATIONS PRACTICE EXAM

10. You are entering a prescription for an albuterol inhaler that delivers 90 mcg per actuation. If each container delivers 200 actuations and each dose is two actuations, how many mg are delivered in each dose?
    a. 180 mg
    b. 90 mg
    c. 0.18 mg
    d. 0.09 mg

11. What is the days supply for a Z-Pak® that contains six azithromycin 250 mg tablets, with directions of 500 mg on the first day, followed by 250 mg once daily until gone?
    a. 3
    b. 5
    c. 4
    d. 6

12. What is the day's supply for metronidazole vaginal gel 70 g, Sig 5 g twice daily?
    a. 5 days
    b. 7 days
    c. 10 days
    d. 14 days

13. What is the days supply for Humulin® N insulin 20 mL, if the dose is 40 U daily?
    a. 100 days
    b. 30 days
    c. 60 days
    d. 50 days

14. How much change should be given to a patient who gives you $50 to pay for 3 prescriptions if the patient's prescription plan has a $10 co-pay?
    a. $20
    b. $10
    c. $40
    d. $30

15. What is the gross profit for a prescription if the selling price is $73.14, the acquisition cost is $52.10, and the AWP is $65.12?
    a. $21.04
    b. $8.02
    c. $13.02
    d. $73.14

16. A prescription has been written for tamsulosin 0.4 mg #100, Sig i cap qd, 2 refills; however, the patient's insurance benefit has a 34-day supply limit. If the original prescription is filled for 34 capsules, how many refills of 34 would be available?
    a. 6 refills of 34
    b. 8 refills of 34
    c. 5 refills of 34
    d. 7 refills of 34

17. You are preparing a prescription for Synthroid® 0.1 mg tablets. How many micrograms are in each tablet?
    a. 0.1 mg
    b. 0.1 mcg
    c. 100 mg
    d. 100 mcg

18. You are entering a prescription for timolol 0.25% ophthalmic solution 5 mL, Sig i gtt o.u. twice daily. How many days should this bottle last if the dropper delivers 20 drops of timolol 0.25% ophthalmic solution per mL?
    a. 10 days
    b. 50 days
    c. 12.5 days
    d. 25 days

19. The dose of a drug is 250 micrograms per kg of body weight. What dose should be given to a child that weighs 55 lbs?
    a. 6.25 micrograms
    b. 6.25 mg
    c. 12.5 mg
    d. 12.5 micrograms

CALCULATIONS PRACTICE EXAM

20. A physician has ordered 15 mL of Brand X antacid suspension to be taken four times daily. How many days will a 12 oz. bottle last?
    a. 6 days
    b. 12 days
    c. 3 days
    d. 10 days

21. How many mcg of digoxin are in 0.4 mL of digoxin solution if the strength of the digoxin solution is 50 mcg per mL
    a. 20 mcg
    b. 125 mcg
    c. 0.02 mcg
    d. 12.5 mcg

22. You are filling a prescription for gentamicin 80 mg. How much gentamicin solution should be measured from a 2 mL vial of gentamicin 40 mg/mL?
    a. 2 mL
    b. 1 mL
    c. 0.8 mL
    d. 1.6 mL

23. 45 units of Humulin® R insulin are to be added to a TPN bag. How much Humulin R (U-100) is needed?
    a. 45 microliters
    b. 0.45 mL
    c. 4.5 mL
    d. 0.45 microliters

24. How much potassium chloride solution (2 mEq/mL) should be added to a 1 liter IV bag if 25 mEq of potassium chloride is needed?
    a. 25 mL
    b. 0.25 mL
    c. 12.5 mL
    d. 50 mL

25. How much sodium chloride is in 25 mL of normal saline?
    a. 225 g
    b. 225 mcg
    c. 225 mg
    d. 2.25 g

26. A drug is in a vial that contains 500 mg of the drug in 2 mL of solution. What is the percent strength of this drug?
    a. 2.5%
    b. 25%
    c. 10%
    d. 5%

27. You have dissolved 20 g of drug in 500 mL of solution. What is the percent strength of the resulting solution?
    a. 0.4%
    b. 4%
    c. 0.2%
    d. 2%

28. How many capsules are needed to prepare 30 mL of 1% clindamycin hydrochloride solution if each capsule contains 150 mg of clindamycin hydrochloride?
    a. 4 capsules
    b. 3 capsules
    c. 1 capsule
    d. 2 capsules

29. How much 1% lidocaine is needed to fill an order for 30 mg of lidocaine?
    a. 0.3 mL
    b. 3 mL
    c. 3 microliters
    d. 0.03 mL

30. What is the percent strength of a 1:100 solution?
    a. 10%
    b. 1%
    c. 0.01%
    d. 0.1%

CALCULATIONS PRACTICE EXAM

31. How much gentian violet is needed to prepare 100 mL of a 1:10,000 solution of gentian violet?
    a. 0.01 mg
    b. 10 g
    c. 10 mg
    d. 10 mcg

32. How much vancomycin should be given per dose for a child that weighs 32 lbs if the dose is 10 mg/kg q6h IV?
    a. 582 mg
    b. 145 mg
    c. 0.69 mg
    d. 704 mg

33. A patient is to receive 100 mg/kg/day of ampicillin. What is the total daily dose for a patient that weighs 40 lbs?
    a. 182 mg
    b. 1.8 g
    c. 18.2 g
    d. 0.182 g

34. If 250 mg of penicillin VK is equivalent to 400,000 Units of penicillin, how many Units of penicillin are in 1 mg of penicillin VK?
    a. 1,600 U
    b. 1,600 MU
    c. 16 MU
    d. 1.6 U

35. A nomogram has been used to determine a patient's BSA is 1.95. If the dose of a drug is 40 mg/sq meter, how much drug should be administered per dose?
    a. 780 mg
    b. 7.8 g
    c. 78 mg
    d. 7.8 mg

36. A 500 mL IV bag is administered over 4 hours. What is the infusion rate?
    a. 125 mL/min
    b. 100 mL/min
    c. 100 mL/hr
    d. 125 mL/hr

37. A 500 mL IV bag is infused at a rate of 100 mL/hr. How long will this bag last?
    a. 2.5 hours
    b. 2 hours
    c. 10 hours
    d. 5 hours

38. An IV has been running at 80 mL/hr for 5 hours and 20 minutes. How much solution has the patient received?
    a. 427 mL
    b. 40 mL
    c. 400 mL
    d. 43 mL

39. An IV is set to deliver 30 drops/min. What is the infusion rate in mL/hr if there are 15 drops/mL?
    a. 30 mL/hr
    b. 40 mL/hr
    c. 120 mLhr
    d. 270 mL/hr

40. How many mL of a 20% solution should be added to 50 mL of a 40% solution to obtain a 25% solution?
    a. 50 mL
    b. 100 mL
    c. 150 mL
    d. 200 mL

41. How many teaspoons equal 20 mL?
    a. 5
    b. 6
    c. 4
    d. 2

CALCULATIONS PRACTICE EXAM

42. If a prescription reads: Aspirin 5 gr, dispense 100 tablets, 1 tablet q 4-6h prn headache, what is the dose in milligrams?
    a. 650
    b. 750
    c. 100
    d. 325

43. How many gallons of Coca Cola™ fountain syrup are needed to package 150 bottles of 120 mL per bottle?
    a. 3
    b. 4
    c. 5
    d. 6

44. If a prescription reads: Amoxicillin 250 mg/5 mL, dispense 150 mL, 375 mg t.i.d. x 5d, what is the dose in household units?
    a. 1 teaspoonful
    b. 1 tablespoonful
    c. 1.5 teaspoonful
    d. 1.5 tablespoonful

45. How many mL are in 2 liters of normal saline?
    a. 200
    b. 2,000
    c. 0.2
    d. 0.002

46. How many mL of KCl 2 mEq/mL are needed if the dose is 30 mEq?
    a. 5
    b. 10
    c. 15
    d. 30

47. How many mL of 25% dextrose are needed to prepare 500 mL of 40% dextrose if you are to prepare 40% dextrose from 25% dextrose and 60% dextrose?
    a. 214 mL
    b. 286 mL
    c. 200 mL
    d. 300 mL

48. If 1 liter is infused over 8 hours, what is the rate of infusion in mL/hr?
    a. 62.5
    b. 100
    c. 125
    d. 250

49. A patient weighs 121 pounds. What is the patient's weight in kg?
    a. 37
    b. 45
    c. 55
    d. 68

50. How many grams of sodium bicarbonate are needed to make 400 mL of a 1:1,000 w/v solution?
    a. 0.2
    b. 0.4
    c. 0.8
    d. 1

TOP 200 MOST-PRESCRIBED DRUGS BY CLASSIFICATION

Following is a list of the top 200 most-prescribed drugs in 2014 grouped by classifications. Since most of these drugs are generic, the generic names are given first and then cross-referenced with corresponding brand names. When a brand name drug is the product in the top 200, it is highlighted in ***bold italics****. This list is based on information in Symphony Health Solutions™* Top 200 Drugs of 2014 *at http://symphonyhealth.com/wp-content/uploads/2015/05/Top-200-Drugs-of-2014.pdf.*

Classification	Generic Name	Brand Name
Analgesic	Fentanyl	Duragesic
Analgesic	Tramadol	Ultram
Analgesic, NSAID	Celecoxib	Celebrex
Analgesic, NSAID	Diclofenac	***Voltaren***
Analgesic, NSAID	Ibuprofen	Motrin
Analgesic, NSAID	Meloxicam	Mobic
Analgesic, NSAID	Naproxen	Naprosyn
Analgesic, Opiate	Acetaminophen with codeine	Tylenol with Codeine
Analgesic, Opiate	Hydrocodone, acetaminophen	Vicodin
Analgesic, Opiate	Morphine sulfate ER	MS Contin
Analgesic, Opiate	Oxycodone	***Oxycontin***
Analgesic, Opiate	Oxycodone, acetaminophen	Percocet
Anesthetic, Local	Lidocaine	Xylocaine
Anti-infective	Amoxicillin	Amoxil
Anti-infective	Amoxicillin, clavulanate	Augmentin
Anti-infective	Azithromycin	Zithromax
Anti-infective	Cefdinir	(Omnicef)
Anti-infective	Cephalexin	Keflex
Anti-infective	Ciprofloxacin	Cipro
Anti-infective	Clindamycin HCl	Cleocin
Anti-infective	Doxycycline hyclate	Doryx
Anti-infective	Levofloxacin	Levaquin
Anti-infective	Metronidazole	Flagyl
Anti-infective	Minocycline HCl	Minocin
Anti-infective	Mupirocin	Bactroban
Anti-infective	Nitrofurantoin mono-macro	Macrobid

**An asterisk next to a drug's classification indicates there are additional common uses for that drug. Parentheses on a brand name drug indicate it is no longer available.*

Classification	Generic Name	Brand Name
Anti-infective	Penicillin V potassium	(V-Cillin K)
Anti-infective	Sulfamethoxazole, trimethoprim	Bactrim
Anti-infective, Antifungal	Fluconazole	Diflucan
Anti-infective, Antifungal	Ketoconazole	Nizoral
Anti-infective, Antifungal	Nystatin	Nilstat
Anti-infective, Antiviral	Acyclovir	Zovirax
Anti-infective, Antiviral	Oseltamivir	***Tamiflu***
Anti-infective, Antiviral	Valacyclovir	Valtrex
Cardiovascular	Carvedilol	Coreg
Cardiovascular	Clopidogrel	Plavix
Cardiovascular	Isosorbide mononitrate ER	(Imdur ER)
Cardiovascular, Anticoagulant	Warfarin	Coumadin
Cardiovascular, Antihyperlipidemic	Atorvastatin	Lipitor
Cardiovascular, Antihyperlipidemic	Ezetimibe	***Zetia***
Cardiovascular, Antihyperlipidemic	Fenofibrate	Tricor
Cardiovascular, Antihyperlipidemic	Gemfibrozil	Lopid
Cardiovascular, Antihyperlipidemic	Lovastatin	Mevacor
Cardiovascular, Antihyperlipidemic	Pravastatin	Pravachol
Cardiovascular, Antihyperlipidemic	Rosuvastatin	***Crestor***
Cardiovascular, Antihyperlipidemic	Simvastatin	Zocor
Cardiovascular, Antihypertensive	Nevibolol	***Bystolic***
Cardiovascular, Antihypertensive*	Amlodipine	Norvasc
Cardiovascular, Antihypertensive*	Amlodipine, benazepril	Lotrel
Cardiovascular, Antihypertensive*	Atenolol	Tenormin
Cardiovascular, Antihypertensive*	Benazepril	Lotensin
Cardiovascular, Antihypertensive*	Clonidine	Catapres
Cardiovascular, Antihypertensive*	Diltiazem	Cardizem
Cardiovascular, Antihypertensive*	Enalapril	Vasotec
Cardiovascular, Antihypertensive*	Furosemide	Lasix
Cardiovascular, Antihypertensive*	Hydralazine	(Apresoline)
Cardiovascular, Antihypertensive*	Hydrochlorothiazide	Microzide
Cardiovascular, Antihypertensive*	Lisinopril	Zestril
Cardiovascular, Antihypertensive*	Lisinopril, hydrochlorothiazide	Zestoretic
Cardiovascular, Antihypertensive*	Losartan	Cozaar
Cardiovascular, Antihypertensive*	Losartan, hydrochlorothiazide	Hyzaar
Cardiovascular, Antihypertensive*	Metoprolol succinate	Toprol

TOP 200 MOST-PRESCRIBED DRUGS BY CLASSIFICATION

Classification	Generic Name	Brand Name
Cardiovascular, Antihypertensive*	Metoprolol tartrate	Lopressor
Cardiovascular, Antihypertensive*	Nifedipine ER	Procardia XL
Cardiovascular, Antihypertensive*	Olmesartan	***Benicar***
Cardiovascular, Antihypertensive*	Propranolol	Inderal
Cardiovascular, Antihypertensive*	Ramipril	Altace
Cardiovascular, Antihypertensive*	Spironolactone	Aldactone
Cardiovascular, Antihypertensive*	Triamterene, hydrochlorothiazide	Dyazide
Cardiovascular, Antihypertensive*	Valsartan	Diovan
Cardiovascular, Antihypertensive*	Valsartan, hydrochlorothiazide	Diovan HCT
Cardiovascular, Antihypertensive*	Verapamil	Calan
Cardiovascular, Antihypertensive*	Rivaroxaban	***Xarelto***
Dental, Mouthwash	Chlorhexidine gluconate	Peridex
Dermatological	Clindamycin phosphate (topical)	Cleocin (Topical)
Dermatological	Clobetasol	Temovate
Dermatological	Clotrimazole, betamethasone	Lotrisone
Dermatological, Antihistamine	Hydroxyzine HCl	(Atarax)
Dermatological, Antihistamine	Hydroxyzine camoate	Vistaril
Electrolyte	Potassium	***Klor-Con***
Electrolyte	Potassium Chloride	Slow-K
Gastrointestinal	Dicyclomine	Bentyl
Gastrointestinal	Phentermine HCl	Adipex-P
Gastrointestinal, Antacid/Anti-GERD*	Dexlansoprazole	***Dexilant***
Gastrointestinal, Antacid/Antiulcer	Esomeprazole	***Nexium***
Gastrointestinal, Antacid/Antiulcer	Famotidine	Pepcid
Gastrointestinal, Antacid/Antiulcer	Lansoprazole	Prevacid
Gastrointestinal, Antacid/Antiulcer	Omeprazole	Prilosec
Gastrointestinal, Antacid/Antiulcer	Pantoprazole	Protonix
Gastrointestinal, Antacid/Antiulcer	Ranitidine	Zantac
Gastrointestinal, Antinausea	Ondansetron	Zofran
Gastrointestinal, Antinausea	Ondansetron ODT	Zofran ODT
Gastrointestinal, Laxative	Polyethylene glycol	Miralax
Hematologic, Hematopoietic	Ferrous sulfate	Feosol

Classification	Generic Name	Brand Name
Hormones & Modifiers, Insulin	Insulin lispro	***Humalog***
Hormones & Modifiers, Adrenal Corticosteroid	Hydrocortisone	Cortef
Hormones & Modifiers, Adrenal Corticosteroid	Methylprednisolone	Medrol
Hormones & Modifiers, Adrenal Corticosteroid	Prednisolone	Orapred
Hormones & Modifiers, Adrenal Corticosteroid	Prednisone	Sterapred
Hormones & Modifiers, Adrenal Corticosteroid	Triamcinolone acetonide	(Aristocort)
Hormones & Modifiers, Contraceptive	Ethinyl estradiol, etonogestrel	***Nuvaring***
Hormones & Modifiers, Estrogen	Conjugated estrogens	***Premarin***
Hormones & Modifiers, Estrogen	Estradiol	Estrace
Hormones & Modifiers, Insulin	Insulin glargine	***Lantus Solostar***
Hormones & Modifiers, Insulin	Insulin glargine (rDNA origin)	Lantus
Hormones & Modifiers, Oral Antidiabetic	Glimepiride	Amaryl
Hormones & Modifiers, Oral Antidiabetic	Glipizide ER	Glucotrol XL
Hormones & Modifiers, Oral Antidiabetic	Glyburide	DiaBeta
Hormones & Modifiers, Oral Antidiabetic	Metformin	Glucophage
Hormones & Modifiers, Oral Antidiabetic	Metformin ER	Glucophage XR
Hormones & Modifiers, Oral Antidiabetic	Pioglitazone	Actos
Hormones & Modifiers, Oral Antidiabetic	Sitagliptin	Januvia
Hormones & Modifiers, Oral Antidiabetic	Ethinyl estradiol, norgestimate	***Tri-Sprintec***
Hormones & Modifiers, Oral Contraceptive	Ethinyl Estradiol, norethindrone, iron	***Microgestin Fe***
Hormones & Modifiers, Oral Contraceptive	Ethinyl Estradiol, norgestimate	***Sprintec***
Hormones & Modifiers, Phosphodiesterase Inhibitor	Sildenafil	Viagra
Hormones & Modifiers, Phosphodiesterase Inhibitor	Tadalafil	Cialis
Hormones & Modifiers, Progestin	Medroxyprogesterone acetate	Provera
Hormones & Modifiers, Thyroid	Desiccated thyroid	***Armour Thyroid***
Hormones & Modifiers, Thyroid	Levothyroxine	***Synthroid***
Musculoskeletal		
Musculoskeletal, Antigout	Allopurinol	Zyloprim
Musculoskeletal, Antirheumatic*	Hydroxychloroquine	Plaquenil
Musculoskeletal, Antirheumatic*	Methotrexate	Rheumatrex
Musculoskeletal, Muscle Relaxant	Baclofen	(Lioresal)
Musculoskeletal, Muscle Relaxant	Carisoprodol	Soma
Musculoskeletal, Muscle Relaxant	Cyclobenzaprine	Flexeril
Musculoskeletal, Muscle Relaxant	Methocarbamol	Robaxin

TOP 200 MOST-PRESCRIBED DRUGS BY CLASSIFICATION

Classification	Generic Name	Brand Name
Musculoskeletal, Muscle Relaxant	Tizanidine	Zanaflex
Musculoskeletal, Osteoporotic	Alendronate	Fosamax
Neurologic (tension headaches)	Butalbital, acetaminophen, caffeine	Fioricet
Neurologic, Antihistamine	Meclizine	Antivert
Neurologic, Antimigraine	Sumatriptan	Imitrex
Neurologic, Antiparkinson	Ropinirole	Requip
Neurological	Lyrica	Pregabalin
Neurological, Antiepileptic*	Gabapentin	Neurontin
Neurological, Antiepileptic*	Lamotrigine	Lamictal
Neurological, Antiepileptic*	Levetiracetam	Keppra
Neurological, Antiepileptic*	Topiramate	Topamax
Ophthalmic, Antiglaucoma	Latanoprost	Xalatan
Psychotropic	Alprazolam	Xanax
Psychotropic	Buprenorphine, naloxone	Suboxone
Psychotropic	Buspirone	(Buspar)
Psychotropic	Citalopram	Celexa
Psychotropic	Clonazepam	Klonopin
Psychotropic	Diazepam	Valium
Psychotropic	Donepezil HCl	Aricept
Psychotropic	Duloxetine	Cymbalta
Psychotropic	Lorazepam	Ativan
Psychotropic	Quetiapine	Seroquel
Psychotropic, Antidepressant	Amitriptyline	(Elavil)
Psychotropic, Antidepressant	Escitalopram	Lexapro
Psychotropic, Antidepressant	Fluoxetine	Prozac
Psychotropic, Antidepressant	Mirtazapine	Remeron
Psychotropic, Antidepressant	Paroxetine	Paxil
Psychotropic, Antidepressant	Sertraline	Zoloft
Psychotropic, Antidepressant	Venlafaxine ER	(Effexor XR)
Psychotropic, Antidepressant*	Bupropion SR	Wellbutrin SR
Psychotropic, Antidepressant*	Bupropion XL	Wellbutrin XL
Psychotropic, Antidepressant*	Trazodone	Desyrel

Classification	Generic Name	Brand Name
Psychotropic, Hypnotic	Temazepam	Restoril
Psychotropic, Hypnotic	Zolpidem	Ambien
Psychotropic, Neurologic, ADHD	Methylphenidate HCl	Ritalin
Psychotropic, Neurologic, ADHD	Amphetamine, dextroamphetamine	Adderall
Psychotropic, Neurologic, ADHD	Dextroamphetamine, amphetamine ER	Adderall XR
Psychotropic,, Neurologic, ADHD	Lisdexamphetamine	***Vyvanse***
Psychotropic, Neurologic, ADHD	Methylphenidate ER	Ritalin SR
Psychotropic. Antipsychotic	Aripiprazole	Abilify
Psychotropic. Antipsychotic	Risperidone	Risperdal
Respiratory	Beclomethasone	***Qvar***
Respiratory	Budesonide, formoterol	Symbicort
Respiratory	Fluticasone	***Flovent***
Respiratory	Fluticasone propionate	Flovent HFA
Respiratory	Fluticasone, salmeterol	***Advair Diskus***
Respiratory	Mometasone	***Nasonex***
Respiratory	Montelukast	Singulair
Respiratory	Tiotropium	Spiriva
Respiratory, Antihistamine	Cetirizine	Zyrtec
Respiratory, Antihistamine	Levocetirizine	Xyzal
Respiratory, Antihistamine	Loratadine	Claritin
Respiratory, Antihistamine	Promethazine	Phenergan
Respiratory, Antitussive	Benzonatate	Tessalon
Respiratory, Bronchodilator	Albuterol	***ProAir HFA***
Respiratory, Bronchodilator	Albuterol	***Proventil HFA***
Respiratory, Bronchodilator	Albuterol	***Ventolin HFA***
Respiratory, Bronchodilator	Albuterol sulfate	(Proventil)
Urinary	Oxybutynin	Ditropan
Urinary*	Doxazosin	Cardura
Urinary*	Finasteride	Proscar
Urinary*	Tamsulosin	Flomax
Vaccine	Influenza vaccine	Fluvirin
Vaccine	Influenza vaccine	***Afluria***
Vaccine	Influenza vaccine	*Fluzone High-Dose*
Vitamin	Cyanocobalamin injection	Vibisone
Vitamin	Folic Acid	Folic Acid
Vitamin	Vitamin D	Vitamin D

COMMONLY REFRIGERATED DRUGS

Many products must be stored at refrigerated temperatures to ensure stability. Following is a list of commonly refrigerated drugs. As per the manufacturer's product information, some products need to be refrigerated immediately upon receipt from the manufacturer. Others may be stored at room temperature but will require refrigeration once reconstituted or when a diluent is added. A few drugs are stored in the refrigerator at the pharmacy but can be kept at room temperature when in use by the patient. For reconstituted products, the manufacturer will provide information as to how long the product is stable once reconstituted.

Brand Name & Dosage Form	Generic Name
ActHib vials	haemophilus b conjugate vaccine (tetanus toxoid conjugate)
Actimmune vials	interferon gamma-1b
Adacel vials	tetanus toxoid, reduced diphtheria toxoid, acellular pertussis vaccine adsorbed (Tdap) vaccine
Alcaine ophthalmic solution	proparacaine
Amoxil suspension	amoxicillin
Anectine vials	succinylcholine chloride
Aspirin Uniserts suppositories	aspirin
Atgam vials	lymphocyte immune globulin
Augmentin suspension	amoxicillin/clavulanic acid
Avonex syringes	interferon beta-1a
Azasite ophthalmic solution	azithromycin
BACiiM vials	bacitracin
Benzamycin gel	erythromycin/benzoyl peroxide
Bicillin syringes	penicillin G benzathine
Cardizem vials	diltiazem hydrochloride
Caverject vials	alprostadil
Ceftin suspension	cefuroxime axetil
Cefzil suspension	cefprozil
Cerebyx vials	fosphenytoin sodium
Cipro suspension	ciprofloxacin
Combipatch transdermal patch	estradiol/norethindrone
Copaxone	glatiramer acetate
DDAVP vials	desmopressin
Digibind vials	digoxin immune Fab
Enbrel syringes	etanercept
Engerix-B vials	hepatitis B vaccine
Epogen vials	epoetin alfa
Genotropin cartridges	somatropin
Havrix vials, syringes	hepatitis A vaccine

Brand Name & Dosage Form	Generic Name
Hemabate vial	carboprost tromethamine
Humalog vials, pens	insulin lispro
Humira syringes	adalimumab
Humulin R vials	regular insulin
Infergen vials	interferon alfacon-1
Kaletra solution, capsules	lopinavir/ritonavir
Lactinex tablets	lactobacillus acidophilus/bulgaricus
Lantus vials, pens	insulin glargine
Leukeran tablets	chlorambucil
Meruvax vials	rubella virus vaccine live
Methergine vials	methylergonovine maleate
Miacalcin nasal spray	calcitonin salmon
MMR II vials	measles, mumps, rubella virus vaccine live
Mycostatin pastilles	nystatin
Neupogen vials, syringes	filgrastim
Nimbex vials	cisatracurium besylate
Norditropin cartridges	somatropin
Norvir soft gels	ritonavir
Neulasta syringe	pegfilgrastim
Neurontin suspension	gabapentin
Pavulon	pancuronium bromide
Phenergan suppositories	promethazine
Pneumovax 23 vaccine	pneumococcal vaccine polyvalent
Premarin Secule vials	conjugated estrogens
RabAvert vial	rabies vaccine
Rapamune solution	sirolimus
Rebetron	ribavirin/interferon alfa-2b
Regranex gel	becaplermin
Risperdal Consta kits	risperidone
Sandostatin vials	octreotide acetate
Survanta vials	beractant
Tamiflu suspension	oseltamivir phosphate
Thyrolar tablets	liotrix
Tracrium vials	atracurium besylate
Veetids suspension	penicillin V
VePesid capsules	etoposide
Vibramycin suspension	doxycycline
Viroptic ophthalmic solution	trifluridine
Xalatan ophthalmic solution	latanoprost
Zemuron vials	rocuronium bromide
Zithromax suspension	azithromycin

ASHP MODEL CURRICULUM GOALS

The ASHP's Model Curriculum for Pharmacy Technician Education and Training Programs *at www.ashp.org/DocLibrary/Accreditation/Model-Curriculum.pdf provides detailed guidance on how to meet the goals defined in ASHP's 2013* Accreditation Standard, *which went into effect in January 2015. There are 45 specific goals grouped into nine goal categories, as outlined below.*

GOALS BY CATEGORY

Personal/Interpersonal Knowledge and Skills

1. Demonstrate ethical conduct in all job-related activities.
2. Present an image appropriate for the profession of pharmacy in appearance and behavior.
3. Communicate clearly when speaking and in writing.
4. Demonstrate a respectful attitude when interacting with diverse patient populations.
5. Apply self-management skills, including time management, stress management, and adapting to change.
6. Apply interpersonal skills, including negotiation skills, conflict resolution, and teamwork.
7. Apply critical thinking skills, creativity, and innovation to solve problems.

Foundational Professional Knowledge and Skills

8. Demonstrate understanding of healthcare occupations and the health care delivery system.
9. Demonstrate understanding of wellness promotion and disease prevention concepts, such as use of health screenings; health practices and environmental factors that impact health; and adverse effects of alcohol, tobacco, and legal and illegal drugs.
10. Demonstrate commitment to excellence in the pharmacy profession and to continuing education and training.
11. Demonstrate knowledge and skills in areas of science relevant to the pharmacy technician's role, including anatomy/physiology and pharmacology.
12. Perform mathematical calculations essential to the duties of pharmacy technicians in a variety of contemporary settings.
13. Demonstrate understanding of the pharmacy technician's role in the medication-use process.
14. Demonstrate understanding of major trends, issues, goals, and initiatives taking place in the pharmacy profession.
15. Demonstrate understanding of nontraditional roles of pharmacy technicians.
16. Identify and describe emerging therapies.

Processing and Handling of Medications and Medication Orders

17. Assist pharmacists in collecting, organizing, and recording demographic and clinical information for direct patient care and medication-use review.
18. Receive and screen prescriptions/medication orders for completeness, accuracy, and authenticity.
19. Assist pharmacists in the identification of patients who desire/require counseling to optimize the use of medications, equipment, and devices.
20. Prepare non-patient-specific medications for distribution (e.g., batch, stock medications).
21. Distribute medications in a manner that follows specified procedures.
22. Practice effective infection control procedures, including preventing transmission of blood borne and airborne diseases.

23. Assist pharmacists in preparing, storing, and distributing medication products requiring special handling and documentation (e.g., controlled substances, immunizations, chemotherapy, investigational drugs, drugs with mandated Risk Evaluation and Mitigation Strategies [REMS]).
24. Assist pharmacists in the monitoring of medication therapy.
25. Prepare patient-specific medications for distribution.
26. Maintain pharmacy facilities and equipment, including automated dispensing equipment.
27. Use material safety data sheets (MSDS) to identify, handle, and safely dispose of hazardous materials.

Sterile and Nonsterile Compounding

28. Prepare medications requiring compounding of sterile products.
29. Prepare medications requiring compounding of nonsterile products.
30. Prepare medications requiring compounding of chemotherapy/hazardous products.

Procurement, Billing, Reimbursement and Inventory Management

31. Initiate, verify, and assist in the adjudication of billing for pharmacy services and goods, and collect payment for these services.
32. Apply accepted procedures in purchasing pharmaceuticals, devices, and supplies.
33. Apply accepted procedures in inventory control of medications, equipment, and devices.
34. Explain pharmacy reimbursement plans for covering pharmacy services.

Patient- and Medication-Safety

35. Apply patient- and medication-safety practices in all aspects of the pharmacy technician's roles.
36. Verify measurements, preparation, and/or packaging of medications produced by other healthcare professionals (e.g., tech-check-tech).
37. Explain pharmacists' roles when they are responding to emergency situations and how pharmacy technicians can assist pharmacists by being certified as Basic Life Support (BLS) Healthcare Providers.
38. Demonstrate skills required for effective emergency preparedness.
39. Assist pharmacists in medication reconciliation.
40. Assist pharmacists in medication therapy management.

Technology and Informatics

41. Describe the use of current technology in the healthcare environment to ensure the safety and accuracy of medication dispensing.

Regulatory Issues

42. Compare and contrast the roles of pharmacists and pharmacy technicians in ensuring pharmacy department compliance with professional standards and relevant legal, regulatory, formulary, contractual, and safety requirements.
43. Maintain confidentiality of patient information.

Quality Assurance

44. Apply quality assurance practices to pharmaceuticals, durable and nondurable medical equipment, devices, and supplies.
45. Explain procedures and communication channels to use in the event of a product recall or shortage, a medication error, or identification of another problem.

ANSWERS TO PRACTICE EXAMS

PTCE PRACTICE EXAM

1. b
2. d
3. a
4. c
5. a
6. c
7. d
8. c
9. b
10. c
11. c
12. d
13. d
14. b
15. c
16. a
17. d
18. c
19. a
20. a
21. a
22. d
23. c
24. b
25. a
26. a
27. c
28. c
29. c
30. c
31. a
32. c
33. a
34. a
35. b
36. d
37. a
38. d
39. b
40. c
41. b
42. a
43. c
44. b
45. c
46. a
47. c
48. b
49. c
50. a
51. b
52. b
53. c
54. b
55. b
56. c
57. c
58. d
59. c
60. c
61. d
62. d
63. c
64. a
65. b
66. b
67. c
68. b
69. c
70. b
71. b
72. a
73. d
74. c
75. b
76. a
77. b
78. b
79. d
80. a
81. b
82. a
83. a
84. c
85. b
86. b
87. c
88. b
89. d
90. d

ExCPT PRACTICE EXAM

1. b
2. b
3. c
4. c
5. b
6. a
7. b
8. b
9. a
10. d
11. d
12. a
13. a
14. a
15. b
16. b
17. d
18. c
19. c
20. d
21. b
22. c
23. c
24. d
25. d
26. d
27. b
28. d
29. c
30. b
31. d
32. a
33. c
34. b
35. d
36. b
37. d
38. b
39. c
40. a
41. a
42. c
43. d
44. b
45. a
46. b
47. b
48. b
49. a
50. b
51. a
52. c
53. c
54. b
55. b
56. c
57. c
58. a
59. c
60. d
61. a
62. c
63. a
64. d
65. d
66. d
67. d
68. a
69. a
70. a
71. b
72. b
73. a
74. d
75. a
76. c
77. b
78. b
79. b
80. a
81. b
82. d
83. a
84. d
85. b
86. b
87. c
88. c
89. d
90. c
91. d
92. c
93. b
94. a
95. c
96. b
97. d
98. d
99. a
100. a
101. a
102. b
103. d
104. c
105. c
106. c
107. a
108. a
109. d
110. d
111. b
112. b
113. d
114. a
115. c
116. c
117. a
118. d
119. d
120. a

CALCULATIONS PRACTICE EXAM

1. c
2. c
3. b
4. a
5. a
6. d
7. d
8. b
9. c
10. c
11. b
12. b
13. d
14. a
15. a
16. d
17. d
18. d
19. b
20. a
21. a
22. a
23. b
24. c
25. c
26. b
27. b
28. d
29. b
30. b
31. c
32. b
33. b
34. a
35. c
36. d
37. d
38. a
39. c
40. c
41. c
42. d
43. c
44. c
45. b
46. c
47. b
48. c
49. c
50. b

ANSWERS TO CHAPTER EXERCISES AND PROBLEMS

CHAPTER 1

p. 4
1. cocaine
2. salicylic acid
3. quinine
4. human genome
5. long-term care
6. pharmacology
7. penicillin
8. MTM services
9. antitoxin
10. antibiotic
11. hormones
12. pharmacognosy
13. Medicare Modernization Act
14. digitalis
15. synthetic

p. 5
1. F
2. T
3. T
4. F
5. T
6. T
7. F
8. F
9. T
10. F

p. 6
1. c
2. d
3. b
4. a
5. c
6. d
7. b
8. a
9. b
10. c

CHAPTER 2

p. 10
1. scope of practice
2. cultural competence
3. confidentiality
4. patient welfare
5. PTCE
6. certification
7. technicians
8. tall man lettering
9. pharmacist
10. performance review
11. ASHP
12. ExCPT
13. continuing education
14. HIPAA

p. 11
1. T
2. F
3. F
4. T
5. T
6. F
7. F
8. T
9. T
10. F

p. 21
1. b
2. d
3. c
4. a
5. b
6. d
7. c
8. b
9. c
10. c

CHAPTER 3

p. 27
1. DEA number
2. beneficence
3. placebo
4. adverse effect
5. legend drug
6. negligence
7. pediatric
8. autonomy
9. Protected Health Information (PHI)
10. liability
11. recall
12. Controlled Substances Act
13. NDC
14. OTC drugs

p. 28
1. F
2. T
3. T
4. T
5. T
6. F
7. T
8. T
9. F
10. T

p. 29
1. Schedule II
2. Schedule III
3. Class III recall
4. Schedule V
5. Class II recall
6. Schedule I
7. Schedule IV
8. Class I recall

p. 30
1. tight, light resistant
2. Qualitest Pharmaceuticals
3. Endocet®
4. oxycodone and acetaminophen
5. tablets
6. oxycodone hydrochloride and acetaminophen
7. C-II
8. room temperature
9. 06/2019

p. 32
1. d
2. a
3. d
4. b
5. a
6. b
7. d
8. c
9. d
10. a

ANSWERS TO CHAPTER EXERCISES AND PROBLEMS

CHAPTER 4

p. 36

1. primary literature
2. tertiary literature
3. *RED BOOK*
4. Safety Data Sheets (SDS)
5. secondary literature
6. Lexicomp
7. *Drug Facts and Comparisons (DFC)*
8. *Physician's Desk Reference*
9. *Handbook on Injectable Drugs*
10. *Kings Guide to Parenteral Admixtures*
11. Martindale
12. the "Orange Book"
13. *Handbook of Nonprescription Drugs*
14. USP–NF

p. 37

1. T
2. T
3. T
4. F
5. F
6. T
7. T
8. T
9. F
10. T
11. T

p. 38

1. c
2. b
3. d
4. d
5. c
6. a
7. d
8. d
9. c
10. a
11. b
12. a
13. c
14. d

CHAPTER 5

p. 47

1. hypotonia
2. fibromyalgia
3. encephalitis
4. endocrine
5. cardiomyopathy
6. parathyroid
7. arthritis
8. hyperglycemia
9. aphagia
10. dyspepsia
11. colitis
12. hypertrophy
13. hernia
14. dysuria
15. phlebotomy
16. thrombosis
17. hematoma
18. hemophilia
19. lymphoma
20. eczema
21. tendinitis
22. neuralgia
23. pachyderm
24. endometriosis
25. vasectomy
26. prostatolith
27. bronchitis
28. phlebitis
29. cystitis
30. lordosis
31. blepharitis

p. 48

1. anorexia
2. aphagia
3. colitis
4. diabetes
5. esophagitis
6. alopecia
7. diaphoresis
8. hyperglycemia
9. bacteremia
10. leukocytes
11. septicemia
12. cellulitis
13. myalgia
14. hypertrophy
15. atrophy
16. paralysis
17. apnea
18. pectoralgia
19. amenorrhea
20. dysuria
21. albuminuria
22. glycosuria
23. hematuria
24. ketouria
25. bradycardia
26. hypertension
27. atherosclerosis

p. 50

1. a
2. b
3. c
4. d
5. c
6. d
7. b
8. c
9. c
10. b
11. c
12. a
13. b

CHAPTER 6

p. 53

1. LXVII
2. XXIX
3. XLI
4. CVIII
5. VI
6. XCVIII
7. IX
8. 19
9. 103
10. 1900
11. 1 ½
12. 20
13. 54

p. 55

1. 0.625
2. 0.002
3. 0.2

Answers to Chapter Exercises and Problems

4. 0.67
5. 0.12
6. 0.385
7. 0.015
8. 70%
9. 75%
10. 150%
11. 25%
12. 4%
13. 80%
14. 2.5%

p. 59
1. Aminosyn® 500 mL
2. dextrose 400 mL
3. KCl 12 mL
4. MVI 5 mL
5. NaCl 5.45 mL
6. sterile water 77.55 mL

p. 60
1. 30
2. 21
3. 3.5
4. 600 mg
5. 300 mL
6. 20 mL
7. 200 mL
8. 7.2 mL
9. 1 mL/min
10. 1.4
11. 180 mL
12. 357 mL of 70% dextrose and 643 mL of sterile water
13. 286 mL 70% dextrose and 714 mL sterile water
14. 250 mL 50% dextrose and 250 mL sterile water
15. 33 mL

Chapter 7

p. 67
1. of each
2. before food, meals
3. right ear or to, up to
4. left ear
5. add water up to
6. left ear
7. each ear
8. twice a day
9. with
10. capsules
11. cream
12. Dispense As Written
13. fluid ounce
14. gram
15. drop
16. bedtime
17. one
18. two
19. intramuscular
20. inhale
21. intravenous
22. iv push
23. iv piggyback
24. liter
25. liquid
26. microgram
27. milliequivalent
28. milligram
29. milliliter
30. no refill
31. right eye
32. left eye
33. left eye
34. each eye
35. after food, after meals
36. by nebulizer
37. afternoon or evening
38. by mouth
39. rectally, into the rectum
40. as needed
41. vaginally, into the vagina
42. each morning
43. every ___ hour(s)
44. four times a day
45. add sufficient quantity to make
46. without
47. subcutaneously
48. slow release
49. sublingually, under the tongue
50. solution
51. one-half
52. immediately
53. suppository
54. suspension
55. syrup
56. three times a day
57. tablet
58. tablespoon
59. topically, locally
60. teaspoon
61. ointment
62. as directed
63. with
64. without

p. 70
Prozac® Prescription
1. Prozac®
2. 20 mg
3. capsules
4. by mouth
5. one capsule every day
6. 2
7. no

Triamcinolone Prescription
1. 0.1%
2. twice daily

p. 71
Protonix® Prescription
1. 30
2. 1

Atrovent® HFA/Flovent® HFA Prescription
1. Each inhaler would last 33 days if two puffs per dose are used or 22 days if three puffs per dose are used. We would normally enter 22 days for the days supply on a prescription of this type.
2. 30 days

p. 72
Synthroid® Prescription
1. DAW

Miacalcin® Prescription
1. Miacalcin® is a nasal

ANSWERS TO CHAPTER EXERCISES AND PROBLEMS

spray. The directions read "one inhalation every day" and the patient should administer the medication to alternate nostrils, meaning that the patient should not use the same nostril two days in a row.

p. 73
Metrogel® Prescription
1. A 70 gram tube should be dispensed
2. "per vagina"

Premarin®/Provera® Prescription
1. Since the patient takes the medication for 21 days and off for 7 days, the medication will last 28 days.
2. Since the patient takes the medication for only 5 days in a 28-day cycle, this medication will also last 28 days.

p. 74
Ortho Novum® 777 Prescription
1. 7 (1 original fill plus 6 refills)

Bactrim® DS Prescription
1. sulfa allergy

p. 75
Cefadroxil Prescription
1. Take one teaspoonful twice daily for 10 days.

Amoxicillin, Biaxin®, and Aciphex® Prescription
1. 28
2. 14
3. 14

p. 76
1. pneumonia, dehydration
2. penicillin allergy
3. every 4-6 hours by mouth
4. 2200 and 600 orders
5. 500 mg by mouth, every 12 hours
6. by mouth, each day

p. 77
1. Lopressor® 50 mg, Hydrochlorothiazide 25 mg, and Sonata® 5 mg
2. one at bedtime as needed

p. 78
Mary Smith Physician Order
1. docusate sodium 100 mg
2. Metamucil®

Andrew Smith Physician Order
1. This medication should be administered as soon as possible.
2. intramuscular injection

p. 79
Steve Smith Physician Order
1. 100 cc per hour
2. hydrochlorothizide 25 mg and Diovan® 80 mg

Barbara Smith Physician Order
1. glyburide 5 mg and Ambien® 5 mg
2. Ambien® 5 mg

p. 80
1. Schedule II drugs
2. extemporaneous compounding
3. formulary
4. product selection code
5. auxiliary label
6. medication order
7. look-alikes
8. Rx
9. DAW
10. high-alert medications
11. transfers
12. DUR
13. NPI
14. DEA number

p. 81
1. T
2. T
3. F
4. T
5. F
6. T
7. T
8. F
9. F
10. F

p. 84
1. c
2. b
3. a
4. b
5. b
6. a
7. c
8. c
9. c
10. c
11. c
12. a
13. b
14. a
15. c

CHAPTER 8

p. 89
1. F
2. T
3. F
4. T
5. F
6. T
7. T
8. T
9. T
10. T

p. 90
1. local effect
2. systemic effect

Answers to Chapter Exercises and Problems

3. buffer system
4. bulk powders
5. buccal
6. hydrates
7. enteric coated
8. water soluble
9. sublingual
10. pH
11. parenteral
12. necrosis
13. ophthalmic
14. intradermal injections
15. intravenous sites
16. hemorrhoid
17. suspensions
18. lacrimal canalicula
19. transcorneal transport
20. emulsion
21. intramuscular injection sites
22. syringeability
23. viscosity
24. aqueous
25. wheal
26. lacrimal gland
27. disintegration
28. colloids
29. Z-tract injection
30. nasal mucosa
31. atomizer
32. nasal inhaler
33. injectability
34. metered dose inhalers
35. percutaneous absorption
36. dissolution
37. syrups
38. sterile
39. Toxic Shock Syndrome
40. diluent
41. IUD

p. 92

1. intraocular
2. intranasal
3. sublingual or oral
4. inhalation
5. peroral
6. intravenous
7. vaginal
8. subcutaneous
9. intramuscular

p. 93, Routes of Administration

1. intradermal
2. subcutaneous
3. intravenous
4. intramuscular

Intramuscular Administration Sites

1. a
2. d
3. b
4. e
5. c

p. 94

1. c
2. a
3. b
4. c
5. c
6. b
7. b
8. a
9. a
10. c
11. a
12. d
13. a
14. d

Chapter 9

p. 104

1. F
2. T
3. T
4. T
5. T
6. F
7. F
8. F
9. F
10. T
11. F
12. T

p. 106

1. extemporaneous compounding
2. USP–NF Chapter <795>
3. USP–NF Chapter <797>
4. calibrate
5. volumetric
6. compounding record
7. meniscus
8. trituration
9. levigation
10. geometric dilution
11. sieves
12. spatulation
13. formulation record
14. aliquot
15. syrup
16. sensitivity
17. flocculating agents
18. thickening agent
19. USP–NF grade
20. ointment
21. emulsifier
22. emulsion
23. nonaqueous solutions
24. beyond-use date
25. hydrophilic emulsifier
26. lipophilic emulsifier
27. primary emulsion
28. mucilage
29. aqueous solutions
30. immiscible
31. pipets

p. 108

1. a
2. b
3. a
4. c
5. b
6. d
7. b
8. b
9. a
10. a
11. a

ANSWERS TO CHAPTER EXERCISES AND PROBLEMS

12. d
13. a
14. b
15. b
16. c

CHAPTER 10

p. 112

1. aseptic techniques
2. pyrogens
3. Flashball
4. isotonic
5. hypertonic
6. hypotonic
7. flow rate
8. heparin lock
9. piggybacks
10. clean rooms
11. admixture
12. buffer capacity
13. diluent
14. anhydrous
15. bevel
16. gauge
17. lumen
18. coring
19. membrane filter
20. compounded sterile preparation
21. final filter
22. laminar flow
23. HEPA filter
24. biological safety cabinets
25. irrigation solution
26. shaft
27. sharps
28. ions
29. osmotic pressure
30. valence
31. lyophilized
32. peritoneal dialysis solution
33. TPN solution
34. primary engineering controls (PECs)
35. dialysis

p. 114

1. F
2. T
3. T
4. F
5. T
6. F

p. 124

1. b
2. c
3. a
4. c
5. b
6. d
7. b
8. c
9. a
10. b
11. d
12. a
13. c
14. b
15. c
16. c

CHAPTER 11

p. 132

1. T
2. T
3. T
4. F
5. F
6. T
7. T
8. F
9. F
10. T
11. F

p. 134

1. biopharmaceutics
2. elimination
3. receptor
4. absorption
5. agonists
6. antagonists
7. complexation
8. duration of action
9. therapeutic window
10. disposition
11. passive diffusion
12. active transport
13. hydrophobic
14. hydrophilic
15. lipoidal
16. gastric emptying time
17. ionized
18. protein binding
19. metabolite
20. enzyme
21. enzyme induction
22. enzyme inhibition
23. first pass metabolism
24. enterohepatic cycling
25. nephron
26. unionized
27. bioavailability
28. bioequivalency
29. pharmaceutical equivalents
30. pharmaceutical alternatives
31. therapeutic equivalents

p. 136

1. b
2. a
3. c
4. b
5. a
6. a
7. c
8. d
9. b
10. d
11. a
12. c
13. a
14. c

CHAPTER 12

p. 140

1. hypersensitivity
2. anaphylactic shock
3. idiosyncrasy
4. hepatotoxicity
5. nephrotoxicity
6. carcinogenicity
7. teratogenicity
8. potentiation
9. pharmacogenomics

ANSWERS TO CHAPTER EXERCISES AND PROBLEMS

10. enzyme inhibition
11. displacement
12. obstructive jaundice
13. antidote
14. hypothyroidism
15. hyperthyroidism
16. complexation
17. cirrhosis
18. adverse drug reaction

p. 141

1. F
2. T
3. T
4. T
5. T
6. T
7. T
8. T
9. T
10. T

p. 142

1. d
2. b
3. c
4. c
5. b
6. d
7. a
8. d
9. a
10. b
11. a
12. b
13. b

CHAPTER 13

p. 152

1. Respiratory Agent
2. Psychotropic Agent
3. Cardiovascular Agent
4. Anti-infective
5. Anti-infective
6. Cardiovascular Agent
7. Cardiovascular Agent
8. Anti-infective
9. Musculoskeletal Agent
10. Cardiovascular Agent
11. Anti-infective
12. Respiratory Agent
13. Anti-infective
14. Psychotropic agent
15. Anti-infective
16. Cardiovascular Agent
17. Dermatological Agent
18. Cardiovascular Agent
19. Gastrointestinal Agent
20. Cardiovascular Agent
21. Psychotropic Agent
22. Anti-infective Agent
23. Respiratory Agent
24. Cardiovascular Agent
25. Antidiabetic
26. Cardiovascular Agent
27. Dermatological Agent
28. Analgesic
29. Antidiabetic Agent
30. Gastrointestinal Agent
31. Anti-infective Agent
32. Cardiovascular Agent
33. Gastrointestinal Agent
34. Cardiovascular Agent
35. Cardiovascular Agent
36. Cardiovascular Agent
37. Gastrointestinal Agent
38. Gastrointestinal Agent
39. Electrolytic Agent
40. Anesthetic
41. Psychotropic Agent
42. Dermatological Agent
43. Electrolytic Agent
44. Antineoplastic
45. Analgesic
46. Gastrointestinal Agent
47. Anti-infective
48. Cardiovascular Agent

p. 154
Analgesics

1. h
2. d
3. a
4. l
5. c
6. i
7. j
8. b
9. f
10. g
11. k
12. e

p. 154
Anti-infectives

1. g
2. c
3. h
4. l
5. b
6. m
7. o
8. q
9. a
10. n
11. k
12. p
13. j
14. r
15. e
16. f
17. s
18. i
19. d

p. 155
Cardiovascular Agents I

1. c
2. a
3. d
4. b
5. g
6. k
7. e
8. h
9. j
10. f
11. i

p. 155
Cardiovascular Agents II

1. a
2. h
3. c
4. i
5. j
6. b

ANSWERS TO CHAPTER EXERCISES AND PROBLEMS

7. n
8. s
9. k
10. t
11. v
12. e
13. o
14. f
15. u
16. d
17. w
18. g
19. r
20. x
21. y
22. q
23. z.
24. p
25. l
26. m

p. 156
Dermatologicals
1. c
2. a
3. d
4. b

p. 157
Gastrointestinal Agents
1. f
2. c
3. a
4. g
5. b
6. j
7. i
8. k
9. h
10. l
11. e
12. d

p. 157
Hormones & Modifiers I
1. c
2. g
3. d
4. h
5. e
6. f
7. b
8. i
9. a

p. 158
Hormones & Modifiers II
1. b
2. e
3. a
4. f
5. c
6. d

p. 158
Hormones & Modifiers III
1. a
2. c
3. d
4. b
5. g
6. e
7. h
8. f

p. 158
Musculoskeletal Agents
1. c
2. d
3. f
4. h
5. g
6. b
7. e
8. a

p. 159
Neurological Agents
1. b
2. a
3. g
4. e
5. f
6. i
7. h
8. d
9. c

p. 159
Psychotropic I
1. c
2. d
3. f
4. h
5. b
6. a
7. i
8. e
9. g

p. 159
Psychotropic II
1. a
2. b
3. d
4. g
5. e
6. h
7. c
8. i
9. f

p. 160
Psychotropic III
1. e
2. a
3. d
4. c
5. b

p. 160
Respiratory Agents I
1. c
2. d
3. f
4. e
5. g
6. b
7. a

p. 160
Respiratory Agents II
1. c
2. f
3. d
4. a
5. e
6. b

ANSWERS TO CHAPTER EXERCISES AND PROBLEMS

p. 161
Urinary Agents
1. d
2. a
3. b
4. c

p. 162
1. T
2. F
3. F
4. T
5. F
6. T
7. F
8. F
9. T
10. F

p. 162
1. c
2. d
3. a
4. b
5. b
6. a
7. c
8. c
9. a
10. b
11. a
12. b
13. a
14. c
15. a

CHAPTER 14

p. 166
1. open formulary
2. closed formulary
3. turnover
4. reverse distributors
5. perpetual inventory
6. drop shipments
7. reorder points
8. MSDS
9. consignment stock
10. purchase order number
11. automated dispensing system
12. unit dose
13. point-of-use stations

p. 167
1. T
2. F
3. T
4. F
5. T
6. F
7. T
8. F
9. F
10. T

p. 168
1. b
2. d
3. b
4. a
5. d
6. b
7. a
8. c
9. d
10. b

CHAPTER 15

p. 172
1. pharmacy benefit managers
2. online adjudication
3. co-insurance
4. co-pay
5. CHAMPUS
6. maximum allowable cost
7. U&C or UCR
8. HMO
9. POS
10. PPO
11. CHAMPVA
12. TRICARE
13. CPT code
14. Medicare
15. Medicaid
16. CMS 1500 form
17. workers' compensation
18. patient assistance programs
19. coordination of benefits

p. 173
1. F
2. T
3. T
4. T
5. T
6. T
7. F
8. T

p.176
1. b
2. a
3. c
4. a
5. d
6. a
7. d
8. c
9. b
10. a

CHAPTER 16

p. 181
1. T
2. T
3. T
4. F
5. F
6. T
7. T
8. T
9. T
10. F
11. T

p. 182
1. red flag rule
2. interpersonal skills
3. walk-in clinics
4. transaction windows
5. partial fills
6. pharmacist's judgement
7. patient profile
8. signature log

ANSWERS TO CHAPTER EXERCISES AND PROBLEMS

9. markup
10. safety caps
11. CMEA
12. shelf stickers
13. disease state management programs
14. auxiliary labels
15. unit price

p. 188
1. b
2. a
3. c
4. c
5. a
6. a
7. c
8. c
9. b
10. b
11. a

CHAPTER 17

p. 195
1. reconstitute
2. CPOE
3. electronic medical record
4. unit dose
5. standing order
6. PRN order
7. STAT order
8. medication administration record (MAR)
9. code cart
10. central pharmacy
11. inpatient pharmacy
12. pharmacy satellite
13. par
14. clean rooms
15. outpatient pharmacy
16. policy and procedure manual
17. standard precautions

p. 196
1. F
2. F
3. T
4. T
5. F
6. T
7. T

p. 202
1. b
2. b
3. d
4. a
5. d
6. a
7. d
8. c
9. c
10. b
11. b
12. b

CHAPTER 18

p. 205
1. F
2. T
3. F
4. T
5. F
6. F
7. T
8. T

p. 206
1. d
2. d
3. d
4. a
5. a
6. c
7. d
8. c
9. a
10. b
11. a

KEY CONCEPTS INDEX

Following is an index to the topics found in the Key Concepts sections of this workbook.

KEY CONCEPTS INDEX

KEY CONCEPTS INDEX

KEY CONCEPTS INDEX

KEY CONCEPTS INDEX